Photovoltaic and Wind Energy Conversion Systems

Photovoltaic and Wind Energy Conversion Systems

Editor

Emilio Figueres

MDPI • Basel • Beijing • Wuhan • Barcelona • Belgrade • Manchester • Tokyo • Cluj • Tianjin

Editor
Emilio Figueres
Universitat Politècnica de
València
Spain

Editorial Office
MDPI
St. Alban-Anlage 66
4052 Basel, Switzerland

This is a reprint of articles from the Special Issue published online in the open access journal *Energies* (ISSN 1996-1073) (available at: https://www.mdpi.com/journal/energies/special_issues/ Photovoltaic_and_Wind_Energy_Conversion_Systems).

For citation purposes, cite each article independently as indicated on the article page online and as indicated below:

LastName, A.A.; LastName, B.B.; LastName, C.C. Article Title. *Journal Name* **Year**, *Volume Number*, Page Range.

ISBN 978-3-0365-0844-3 (Hbk)
ISBN 978-3-0365-0845-0 (PDF)

Contents

About the Editor

Emilio Figueres received his M.Sc. degree from the Ecole Nationale Supérieure d'Electrotechnique, d'Electronique, d'Informatiqueet d'Hydraulique de Toulouse, Toulouse, France, in1995, and his Dr. Ingeniero Industrial (Ph.D.) degree from the Universidad Politécnica de Valencia (UPV), Valencia, Spain, in 2001. Since 1996, he has been associated with the Electronics Engineering Department, UPV, where he was Head of the Department from 2008 to 2016, and currently works as a Full Professor. His main research interests include modeling and control of power converters, power processing of renewable energy sources, and grid-connected converters for distributed power generation and improvement of power quality. In the above areas, he has co-authored over 100 papers published in Journals Citation Report (JCR) indexed journals and conferences. He habitually collaborates with companies such as Power Electronics, Mahle Electronics, Ingeteam Power Technology, among others, in the development of power converters for processing of renewable energies, onboard chargers for electric vehicles, and auxiliary systems in railway applications. He holds several patents on these topics.

Preface to "Photovoltaic and Wind Energy Conversion Systems"

In the barely two decades since the advent of the 21st century, renewable energies have gone from being one of the great opportunities for the future of humanity to become one of the main players on the world electricity production stage. Except for hydraulics, the two types of renewable energy that have contributed to this radical change the most have been solar and wind energies. Indeed, according to the statistics published by the International Renewable Energy Agency (IRENA) in 2018 (the latest with data available at this time), hydropower produced 4,149,215 GWh (63% of the total production of all energies renewable energy), onshore wind produced 1,194,718 GWh (18.1%), and solar power contributed with 549,833 GWh (8.3%), which clearly shows the importance of both technologies, especially remarkable in their increasing weight in the production of electricity at a global scale. In this context, this Special Issue of Energies highlights some of the latest technological advances in solar photovoltaic and wind technologies, including maximum power point tracking algorithms, parallel connection of central inverters in high power photovoltaic plants, modeling of systems, etc. Likewise, there is also the participation of authors who show a current overview of the state-of-the-art in solar concentrators and also applied case studies.

The authors of the papers and I hope that readers will find the contributions interesting and useful for their personal and professional development, or that the papers simply allow them to delve into some aspects of two technologies that are as exciting as they are relevant to the future of humanity.

Emilio Figueres
Editor

Article

A Control Scheme without Sensors at the PV Source for Cost and Size Reduction in Two-Stage Grid Connected Inverters

Raúl González-Medina *, Marian Liberos, Silvia Marzal, Emilio Figueres and Gabriel Garcerá

Grupo de Sistemas Electrónicos Industriales del Departamento de Ingeniería Electrónica, Universitat Politècnica de València, Camino de Vera s/n, 46022 Valencia, Spain
* Correspondence: raugonme@upv.es; Tel.: +34-963-879-606

Received: 21 June 2019; Accepted: 23 July 2019; Published: 1 August 2019

Abstract: In order to reduce the cost of PV facilities, the market requires low cost and highly reliable PV inverters, which must comply with several regulations. Some research has focused on decreasing the distortion of the current injected into the grid, reducing the size of the DC-link capacitors and removing sensors, while keeping a good performance of the maximum power point tracking (MPPT) algorithms. Although those objectives are different, all of them are linked to the inverter DC-link voltage control loop. Both the reduction of the DC-link capacitance and the use of sensorless MPPT algorithms require a voltage control loop faster than that of conventional implementations in order to perform properly, but the distortion of the current injected into the grid might rise as a result. This research studies a complete solution for two-stage grid-connected PV inverters, based on the features of second-order generalized integrators. The experimental tests show that the proposed implementation has a performance similar to that of the conventional control of two-stage PV inverters but at a much lower cost.

Keywords: photovoltaics; two-stage grid-connected PV inverters; reduced DC-link; sensorless MPPT

---

1. Introduction

In order to reduce the installation and maintenance costs, the photovoltaic (PV) market requires low cost and reliable systems. Moreover, grid-connected PV inverters must comply with several electromagnetic compatibility (EMC) regulations, some of which limit the distortion of the current injected into the grid (THDi), like IEC EN61000 and IEEE519 [1–4]. Maximum power point tracking (MPPT) algorithms are implemented to optimize the performance of PV systems [5–9]. Conventional MPPT algorithms use current and voltage sensors to calculate the power extracted from the PV source. Several investigations have focused on reducing the number of sensors in PV inverters when implementing MPPT algorithms [10–13], which has a positive impact on cost reduction.

One possible solution to achieve both a good maximum point tracking (MPPT) performance and a reduced THDi is the use of two-stage grid-connected PV inverters, based on a DC-DC converter connected to the PV source, followed by a grid-connected inverter [14–16]. This paper focused on this kind of topologies.

There is a trend to reduce the required capacitance at the DC-link between the DC-DC converter and the inverter stage, which allows replacing electrolytic capacitors by film capacitors, which are more durable [17–22]. Two major effects of the DC-link capacitance reduction are the increase of the voltage ripple at the capacitors and higher transient variations of that voltage under dynamic operation point changes of the inverter. These variations must be bound for the proper operation of the inverter.

In this research, the implementation of control structures based on second-order generalized integrators (SOGI) [23–27] is proposed to support the DC-link capacitance reduction of two-stage grid-connected PV inverters with a small number of sensors. On the one hand, it will be shown that SOGIs can improve the dynamic response of the DC-link voltage. On the other hand, the frequency adaptability of SOGI structures will help to reduce the THDi in the case of variations of the grid frequency, even under highly distorted grid voltage conditions.

A sensorless MPPT algorithm that does not require sensors of the PV panel electrical variables was developed in this study. It is based on the power balance at the DC-link [11] and the improvement of the PV inverter voltage control loop achieved by the use of SOGIs.

The applied techniques allow both a reduced DC-link and a sensorless MPPT algorithm, keeping the performance similar to that of conventional MPPTs, but at a much lower cost.

2. Two-Stage Grid-Connected PV Inverter

The two-stage grid-connected PV inverter shown in Figure 1 has been used for validating the theoretical study. It is formed by a flyback DC-DC and a single phase inverter. The inverter connects a single 230 W PV panel to the single-phase grid (230 Vrms, 50 Hz).

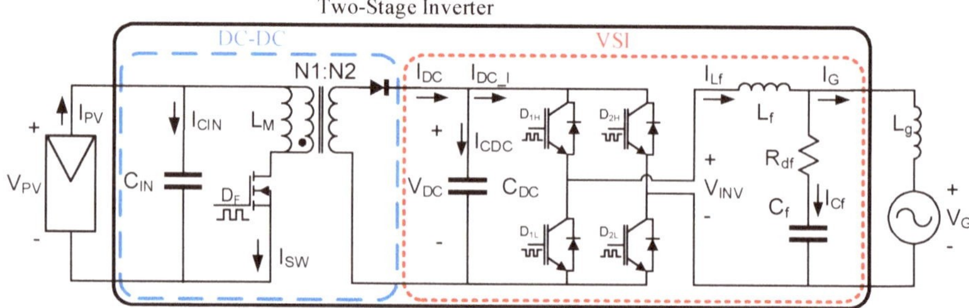

Figure 1. Two-stage grid-connected inverter.

The inverter has a reduced size DC-link. Voltage and current sensors for the measurement of the PV panel voltage, V_{PV}, and current, I_{PV}, are available, in order to compare conventional MPPT algorithms, which make use of those sensors, with sensorless MPPT algorithms under the same conditions.

Costs saving associated with a smaller bulk capacitor and to the absence of MPPT sensors depends on several characteristics, but the estimation done for the implemented prototype is detailed in Table A1 of Appendix A.

2.1. Grid-tied VSI

The voltage source inverter (VSI) is formed by the DC-link capacitance (C_{DC}), a full bridge of IGBTs and an LCL grid filter. The full-bridge is commutated by means of unipolar sinusoidal pulse width modulation (SPWM). The grid filter was designed following the guidelines of Reference [28]. The values of the inverter are shown in Table 1.

The current I_{CDC} has high-frequency current components (I_{DC_SW}) due to the switching of the transistors and a low-frequency current component (I_{DC_AC}). The frequency of I_{DC_AC} is twice the grid frequency and causes a voltage ripple at the DC-link (V_{DC_R}). In Reference [18] it was shown that the minimum value of C_{DC} is determined by the maximum and minimum permissible voltage at the DC-link caused by V_{DC_R}. However, this criterion does not consider the dynamics of the DC-link [19] and transient voltage variations, that are due to changes in the operation point. The proposed value of the capacitance, C_{DC}, to study the effects of a reduced DC-link is designed to limit the peak to peak value of V_{DC_R} to 10% of V_{DC} at nominal power (P_G = 230 W), thus the value of the capacitance C_{DC} is calculated following Equation (1).

$$C_{DC} \geq \sqrt{2} \cdot \frac{V_{G_{RMS}} \cdot I_{G_{RMS}}}{V_{DC_{AV}} \cdot \pi \cdot F_{RDC} \cdot V_{DC_R\ PP}} = \frac{230V \cdot 1A}{380V \cdot \pi \cdot 100Hz \cdot 38V} = 50.7\ \mu F \approx 50\ \mu F \tag{1}$$

Table 1. Values of the Voltage Source Inverter (VSI) stage.

Item	Value
Topology	Single phase, full-bridge
RMS Grid Voltage: V_G	230 V_{AC}
Grid frequency: F_G	50 Hz
Rated Output Power: P_G	230 W
Max Current: I_G	1 A
Modulation	Unipolar SPWM
Inverter switching Frequency: F_{SW_I}	20 kHz
Sampling Frequency: F_S	40 kHz
Filter Inductance: L_f	38 mH
Filter Capacitance: Cf	330 nF
Damping Resistor: Rf	50 Ω
Grid Inductance estimation: L_g	1.5 mH (strong grid) 3 mH (standard grid) 6 mH (weak grid)
DC-link Voltage: V_{DC}	380 V
DC-link Voltage Ripple: V_{DC_R}	10% of V_{DC} (38 $V_{pk\text{-}pk}$)
DC-link Capacitance: C_{DC}	50 μF

2.2. Step Up DC-DC Converter

The DC-DC stage, shown in Figure 1, is a Flyback converter designed for boosting the voltage from the PV panel (V_{PV}) up to the voltage of the DC-link (V_{DC}) and providing high-frequency galvanic isolation between the PV panel and the grid. The converter is designed to work in discontinuous conduction mode (DCM) because the value of the transformer magnetizing inductance (L_M) and the physical size of the transformer become smaller [15]. The MPPT algorithm establishes the operation point of this stage since the PV panel voltage is at the input of the DC-DC converter. It is worth pointing out that in the two-stage PV inverter structure the output voltage of the DC-DC converter is regulated by the inverter stage, whereas the panel voltage is controlled by the DC-DC converter following the reference value provided by the MPPT algorithm.

The current I_{SW}, through the switch of the DC-DC converter, is composed by Equation (2) an average value equal to the PV panel current, I_{PV}, and a high-frequency component, I_{CIN}, provided by a low voltage input capacitance, C_{IN}, following Equation (2).

$$I_{SW} = I_{PV} - I_{CIN}. \tag{2}$$

The size of C_{IN} depends on the high-frequency current components at the input of the DC-DC converter. Besides, the value of the output capacitance, C_{DC}, has an influence on the MPPT performance, since V_{PV} is susceptible to the low-frequency voltage ripple at the DC-link voltage, V_{DC_R}. It is worth pointing out that the implementation of a peak current control (PCC) is highly desirable for protecting the power switches from transient overcurrents. The values of the DC-DC converter are detailed in Table 2.

Table 2. Values of the DC-DC stage.

Item	Value
Topology	Flyback
DC Input Voltage: V_{PV}	24 V to 35 V at the MPPT
DC Output Voltage: V_{DC}	380 V
Rated Input Power: P_{PV}	230 W
Max Input Current: I_{PV}	8 A
Flyback converter switching Frequency: F_{SW_F}	24 kHz
Input Capacitance: C_{IN}	4 mF
Transformer Turns Ratio: N = N1/N2	1/16
Transformer Magnetizing Inductance: L_M	10 μH
Conduction Mode	Discontinuous (DCM)

3. Control

The control of the two-stage inverter has been implemented digitally in a Texas Instruments TMS320F28335 [29] microcontroller with digital signal processor (DSP) extensions at a sampling frequency (F_S) of 40 kHz. The controllers have been calculated in the continuous domain, having taken into account the digital delays, and then discretized using the bilinear "Tustin" transform. The delay between the sampling and the update of the reference inside the DSP has been done by using a second-order Padé approximation.

It is worth pointing out that the dynamic models used in this control study result from perturbing the averaged variables of the DC-DC converter or of the inverter stage around an operation point, as expressed by Equation (3). In Equation (3), X and $\hat{x}$ denote the operation point value and the small-signal term of the averaged variable, x, respectively. The averaging is done in every cycle of the switching frequency.

$$x = X + \hat{x} \tag{3}$$

3.1. Control Scheme of the VSI Stage

The complete control structure of the VSI is shown in Figure 2.

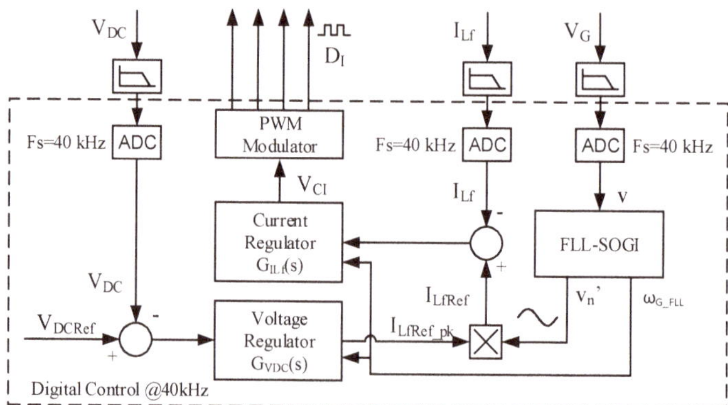

Figure 2. VSI control scheme.

3.1.1. Synchronization with the Grid

The synchronization with the grid voltage (V_G) has been implemented by means of an SOGI based Frequency Locked Loop (FLL-SOGI) [27], which provides the calculation of the grid frequency in rad/s, ω_{G_FLL}, a sinusoidal signal in phase with the grid, v', and a sinusoidal signal in quadrature, qv'. The amplitude of the fundamental harmonic of V_G, v'_{pk}, is calculated following Equation (4), and a normalized sinusoidal signal in phase with the grid, v'_n, is obtained in Equation (5).

$$v'_{pk} = \sqrt{v'^2 + qv'^2} \tag{4}$$

$$v'_n = \frac{v'}{v'_{pk}} \rightarrow v'_n(t) \equiv \cos(\omega_{G\_FLL} \cdot t) \tag{5}$$

3.1.2. Control of the Current Injected into the Grid

The control of the current injected into the grid, I_G, is indirectly performed by controlling the current through the inductance Lf of the filter, I_{Lf}, because the control of I_{Lf} is less sensitive to grid impedance variations [30].

A proportional + resonant controller + a harmonics compensator (P + R + HC) current regulator, $G_{ILf}(s)$, expressed by Equation (6), has been designed following Reference [26] and References [31–34] for tracking the sinusoidal reference of I_{Lf}, $I_{Lf\,Ref}$. Both the resonant and the harmonics compensator have been implemented by means of second-order generalized integrators (SOGI).

$$G_{ILf}(s) = P_{Lf} + R_{Lf}(s) + HC(s) = K_{PLf} + \sum_{i=1,3,5,7} K_{RLf[i]} \frac{K_{BWRLf[i]} \cdot (\omega_{G\_FLL} \cdot i) \cdot s}{s^2 + K_{BWRLf[i]} \cdot (\omega_{G\_FLL} \cdot i) \cdot s + (\omega_{G\_FLL} \cdot i)^2} \tag{6}$$

Taking into account that the value of ω_{G_FLL} used in Equation (6), provided by the FLL-SOGI, it can be concluded that G_{ILf} is adaptive in frequency, allowing high performance even under large variations of the grid frequency. The index 'i' in Equation (6) represents the corresponding harmonic. The gains of G_{ILf} are shown in Table 3.

Table 3. Constants of the regulator P + R + HC.

Constant	Value
K_{PLf}	0.65
K_{RLf1}	100
K_{BWRLf1}	0.02
K_{RLf3}	100
K_{BWRLf3}	0.02/3
K_{RLf5}	100
K_{BWRLf5}	0.02/5
K_{RLf7}	25
K_{BWRLf7}	0.02/7

3.1.3. Control of the DC-link Voltage (V_{DC})

A reduced DC-link capacitance leads to the fast dynamics of the V_{DC} control loop at the expense of an increase of the THDi of I_G [16,17]. In Reference [16] a notch filter in the DC-link voltage control loop was implemented to reduce the low-frequency harmonics of I_G. In the current study, the notch filter is implemented by means of SOGIs, achieving adaptation to grid frequency variations. This implementation allows an increase of the crossover frequency of the V_{DC} control loop without increasing the distortion of I_G, even with a high low voltage ripple at the DC-link and under large grid frequency variations. The control scheme of V_{DC} is shown in Figure 3.

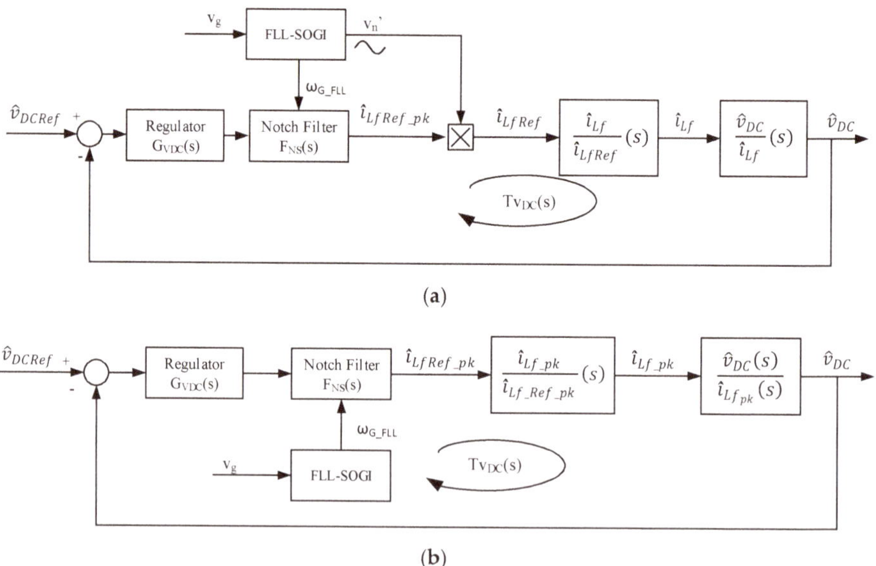

(a)

(b)

Figure 3. Control loop of the DC-link voltage (V_{DC}) with second order generalized integrator (SOGI) notch filter: (**a**) detailed model, (**b**) equivalent model.

In Figure 3, two small-signal transfer functions play an important role. The first one is the transfer function $\hat{i}_{Lf}/\hat{i}_{LfRef}(s)$ in Figure 3a, which is obtained by closing the control loop of I_{Lf}. This transfer function can be approximated by (7) in Figure 3b, where ω_{Ci} is the crossover frequency of the VSI current loop.

$$\frac{\hat{i}_{Lf\_pk}}{\hat{i}_{Lf\_Ref\_pk}}(s) \approx \frac{1}{1 + \frac{s}{\omega_{Ci}}} \tag{7}$$

In the case of a three-phase grid connected inverter, the derivation of the transfer function from the AC side active current to the DC-link voltage and the adjustment of the voltage loop PI regulator can be found in Reference [35], pages 210–219. In the case of the single-phase inverter under study, an analogous Equation (8) can be derived, based on the power balance and the power perturbation at the DC and AC sides. The transfer function from the peak value of the inverter output current at the AC side to the DC-link voltage can be expressed by Equation (8) after some derivation. Note that Equation (8) consists of a first-order transfer function with an (unstable) right half plane (RHP) pole, ω_{P_RHP}, whose value depends on the operation point values V_{DC} and I_{DC}.

$$\frac{\hat{v}_{DC}(s)}{\hat{i}_{Lf\_pk}(s)} = \frac{\frac{V_G}{\sqrt{2}\cdot I_{DC}}}{1 - \frac{s}{\left(\frac{I_{DC}}{C_{DC}\cdot V_{DC}}\right)}} = \frac{\frac{V_G}{\sqrt{2}\cdot I_{DC}}}{1 - \frac{s}{\omega_{P\_RHP}}}; \quad \omega_{P\_RHP} = \frac{I_{DC}}{C_{DC}\cdot V_{DC}} \tag{8}$$

The loop gain $Tv_{DC}(s)$ of Figure 3 is tuned by means of the PI regulator $G_{VDC}(s)$. Equation (9) provides the crossover frequency, $Fc_{\_VDC}$, one decade higher than $\omega_{P\_RHP}/(2\pi)$. Note that an open loop unstable system can be stabilized by feedback only if the loop gain has a gain crossover frequency much higher than the maximum possible value of the unstable open loop pole, $Fc_{\_VDC} \gg \omega_{P\_RHP}/(2\pi)$ in this case. Besides, the value of $Fc_{\_VDC}$ must be much lower than twice the grid frequency (F_G), to reduce the effect of the low-frequency voltage ripple at the DC-link (f_{ripple} = 2 F_{c_VDC}) in the current reference signal i_{Lf_ref}, which could produce an unacceptable distortion of the grid injected current.

$$Gv_{DC}(s) = -0.03902 \cdot \left(\frac{s + 0.6283}{s} \right) \tag{9}$$

It can be observed from Figure 3 that a notch filter, $F_{NS}(s)$, is placed in series with the PI controller $Gv_{DC}(s)$. The expression of the notch filter transfer function is given by Equation (10). The center frequency of $F_{NS}(s)$ is twice the grid frequency (ω_{NS} = 2 ω_{G_FLL}) in order to filter the ripple at f_{ripple} coming from the sensed DC-link voltage. The tuning of the notch filter is provided by the FLL-SOGI previously described. The constant K_{NS} is used to adjust the bandwidth of the filter, BW_{NS}, as shown in Equation (11). The notch filter allows getting a high enough crossover frequency of the voltage loop with no distortion of the grid injected current. Note that a fast enough DC-link voltage loop is crucial to keep the DC-link voltage within safe values in reduced size DC-links with low capacitance.

$$F_{NS}(s) = \frac{s^2 + \omega_{NS}^2}{s^2 + K_{NS} \cdot \omega_{NS} \cdot s + \omega_{NS}^2} \tag{10}$$

$$BW_{NS} \cdot 2\pi = K_{NS} \cdot \omega_{NS} \rightarrow K_{NS} = \frac{2\pi \cdot BW_{NS}}{\omega_{NS}} = \frac{100 \cdot 2\pi}{2 \cdot F_G \cdot 2\pi} = 1 \tag{11}$$

The Bode plots of $T_{VDC}(s)$ depicted in Figure 4a are those obtained when the notch filter, $F_{NS}(s)$, placed in series with $Gv_{DC}(s)$ isn't used. The PI regulator $G_{VDC}(s)$ (9) has been tuned in order to achieve a crossover frequency F_{C_VDC} = 53 Hz, with a phase margin higher than 82° (PM > 82°) and a gain margin higher than 70 dB (GM > 70 dB). The system is stable but the attenuation at 100 Hz is just 5.5 dB. Therefore, the output of the voltage regulator has a remarkable low-frequency voltage ripple due to V_{DC_R}, thus producing a high distortion of the grid current.

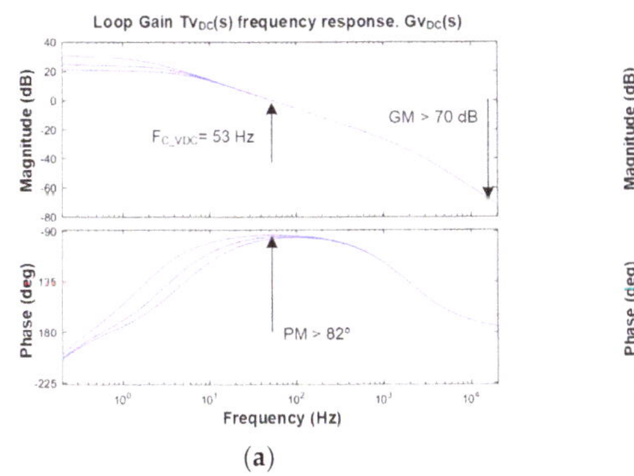

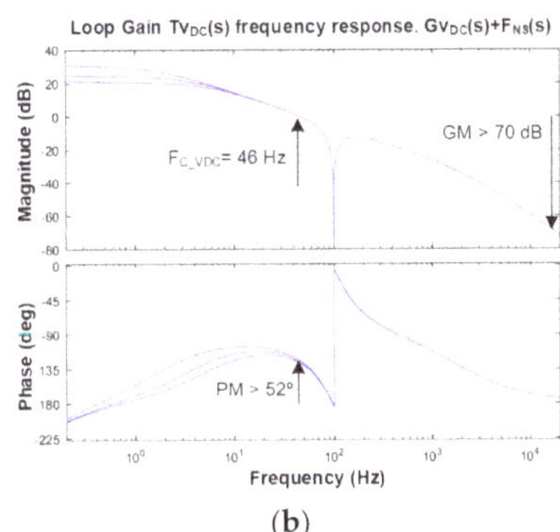

Figure 4. Loop gain frequency response of $T_{vDC}(s)$ @ I_{DC} = [0.2, 0.4, 0.6] A. C_{DC} = 50 µF: (**a**) without notch SOGI filter, (**b**) with notch SOGI filter.

Figure 4b shows the Bode plots of $T_{VDC}(s)$ when the notch filter is used in series with $G_{VDC}(s)$. In that case the value of F_{C_VDC} has slightly decreased (F_{C_VDC} = 46 Hz), getting high stability margins: PM > 52° and GM > 70 dB. The system is also stable, but the attenuation at 100 Hz is higher than 100 dB.

3.2. Control Scheme of the DC-DC Stage

The control structure of the DC-DC stage is depicted in Figure 5. It is composed by an outer digital voltage loop, regulating V_{PV}, in cascade with an analog peak current control (PCC) circuit, which sets the peak value of the current, I_{SW}, through the Flyback converter power transistor.

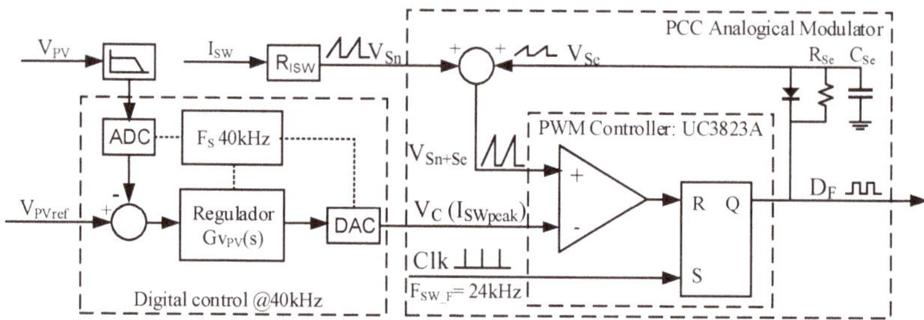

Figure 5. Flyback DC-DC PCC control scheme.

3.2.1. Peak Current Control of DC-DC Stage

The PCC control scheme shown in Figure 5 and has been designed following Reference [15]. This control structure is based on the cycle-by-cycle measurement of the current through the transistor, I_{SW}, of the DC-DC converter. The peak value of I_{SW} is limited by the control signal V_C. An external stabilization ramp signal, V_{Se}, is added to the sensed current signal, V_{Sn}. This method also provides protection for both the HF transformer and the power transistors against an eventual overcurrent.

The modulation index of the PCC, mc = 1 + Se/Sn (Equation (13)), is tuned by means of the slope Se of the external ramp V_{Se}. The value Se = 110 V/ms accomplishes a dynamic behaviour of $\hat{v}_{V_{PV}}/\hat{v}_C(s)$ close to that of a first-order system, as it can be observed from Figure 6a. The high V_{DC_R} ripple value produced by the low size of C_{DC} can change the operation point along the I–V curve of the PV source, degrading the MPPT performance. Therefore, a low susceptibility of V_{PV} to the ripple V_{DC_R} is required. The open loop susceptibility of V_{PV} to variations of V_{DC} (Equation (14)) at 100 Hz is lower than −41.5dB as shown in Figure 6b, therefore V_{DC_R}, that is 10% of V_{DC} (40 Vpp), causes a V_{PV} voltage ripple of 340 mVpp. It can be concluded that the 100 Hz ripple at V_{DC} has a low influence on the PV voltage. Therefore, the sensing of V_{PV} could be avoided and still a good MPPT would be obtained.

$$S_N = R_i \frac{V_{PV}}{L_m} \tag{12}$$

$$F_M = \frac{1}{(S_n + S_e) \cdot T_{SW\_F}} = \frac{1}{m_c \cdot S_n \cdot T_{SW\_F}} \tag{13}$$

$$A(s) = \left. \frac{\hat{v}_{V_{PV}}(s)}{\hat{v}_{DC}(s)} \right|_{\hat{v}_C=0} \tag{14}$$

3.2.2. PV Panel Voltage (V_{PV}) Control Loop in the Conventional MPPT

The reference of V_{PV} (V_{PVRef}) is updated at the sampling frequency of the MPPT, F_{MPPT}. The PV panel voltage control loop is implemented digitally and its control scheme is depicted in Figure 7. The sampling frequency of the control loop (Fs) is 40 kHz. This control loop is adjusted by means the PI regulator $G_{VPV}(s)$, whose values are shown by Equation (15). The crossover frequency, Fc_{VPV}, of the loop gain $T_{VPV}(s)$ must be much higher than F_{MPPT} so that V_{PV} can track V_{PVref}. The transfer function $\hat{v}_{V_{PV}}/\hat{v}_C(s)$ is the closed loop of the PCC and $\hat{v}_{V_{PV}}/\hat{v}_C(s)$ is the open loop susceptibility of V_{PV} to the variations of V_{DC}.

As it can be observed from the Bode plots of the loop $T_{VPV}(s)$ in Figure 8a, the crossover frequency Fc_{VPV} achieved by $G_{VPV}(s)$ is higher than 100 Hz. Therefore, an MPPT algorithm running at F_{MPPT} = 10 Hz is suitable Figure 8b shows that the presence of this control loop reduces the susceptibility $\hat{v}_{V_{PV}}/\hat{v}_{DC}(s)$ (16) at 100 Hz down to −55 dB, therefore the voltage ripple in V_{PV} caused by V_{DC_R} is 71 mVpp. Note that the use of a control loop of V_{PV} reduces the sensitivity of V_{PV} to the 100 Hz ripple in V_{DC} in a factor of around 5: The ripple in V_{PV} is 71 mVpp with voltage loop compared with 340 mVpp with only PCC loop.

$$G_{VPV}(s) = -1.8345\left(\frac{s+350}{s}\right) \tag{15}$$

$$A_{CL}(s) = \left.\frac{\hat{v}_{PV}(s)}{\hat{v}_{DC}(s)}\right|_{\hat{v}_{PVRef}=0} \tag{16}$$

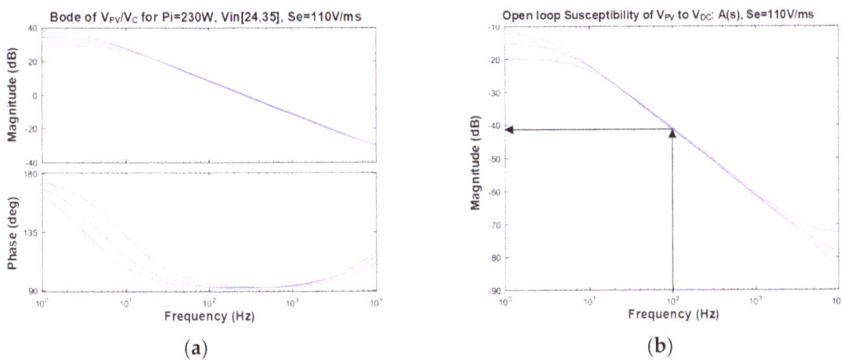

Figure 6. Frequency response of the PCC: (**a**) closed loop $\hat{v}_{V_{PV}}/\hat{v}_C(s)$ response and (**b**) open loop susceptibility: A(s). $V_{PV}\in[24, 30, 35]$ V, $P_{PV} = 230$ W, Se = 110 V/ms.

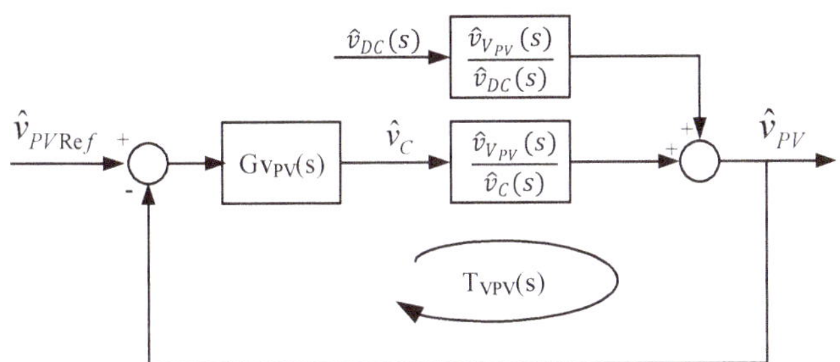

Figure 7. V_{PV} control loop.

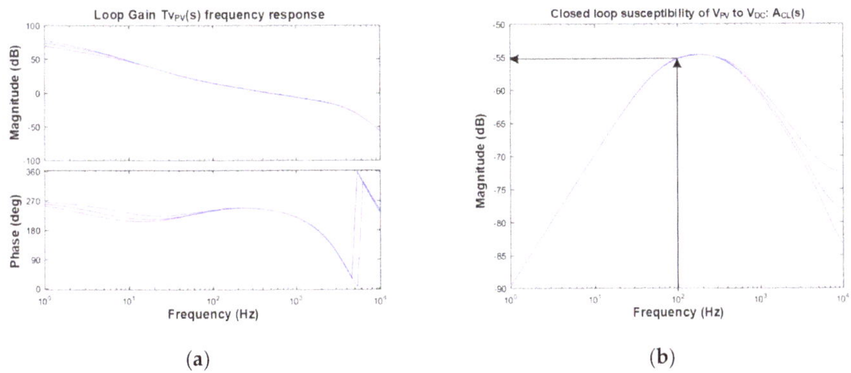

Figure 8. Frequency response of V_{PV} control loop: (**a**) $T_{VPV}(s)$ loop gain of the V_{PV} control loop and (**b**) closed loop susceptibility, $A_{CL}(s)$. $V_{PV} = [24, 30, 35]$ V, $P_{PV} = 230$ W, Se = 110 V/ms.

4. MPPT Implementation without V_{PV} and I_{PV} Sensors

The conventional implementations of MMPT algorithms use current and voltage sensors to measure the voltage (V_{PV}) and the current (I_{PV}) of the PV source as it is shown in Figure 9. The use of voltage and current sensors increases the cost of the power converter. Sensorless MPPT algorithms have been developed [10–13] in order to reduce the number of sensors, yielding a cost reduction. In [11] a sensorless MPPT implementation based on the power balance at the DC-link was presented, which relies on the fact that the power injected into the grid (P_G) can be considered almost equal to the power extracted from the PV panels, $P_G \approx P_{PV}$. In that implementation, the reference of the current injected into the grid (I_G) is used as an estimation of P_G. This reference current depends on the control loop of V_{DC} so that the sampling frequency of the MPPT algorithm (F_{MPPT}) is limited by the dynamics of that loop. Besides, the reference of I_G is sensitive to the variations of the grid voltage and has a low-frequency ripple due to V_{DC_R}. Moreover, the method explained in Reference [11] is based on the assumption that the amplitude of the grid voltage is stable.

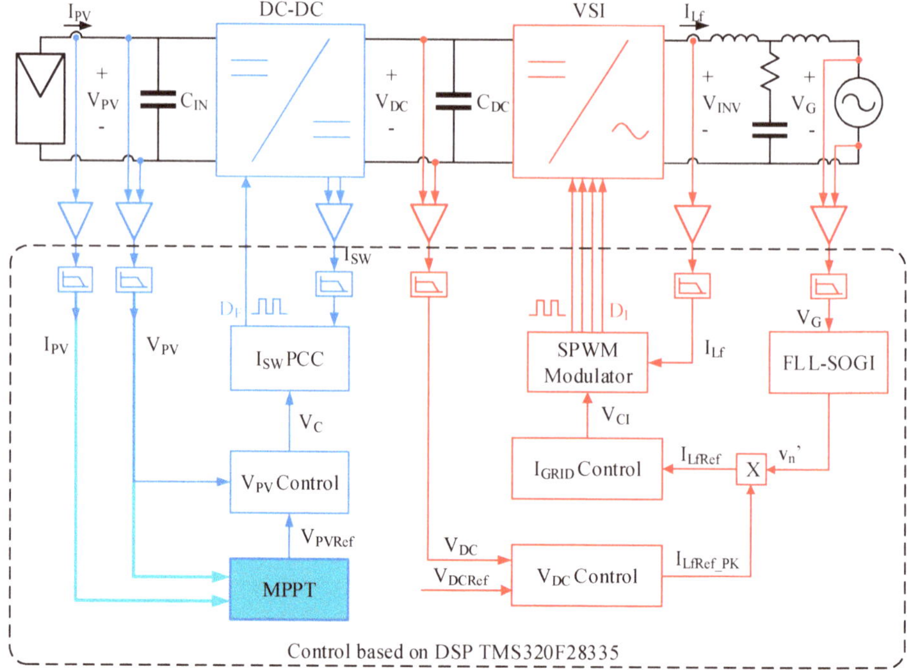

Figure 9. Conventional MPPT implementation.

A sensorless MPPT implementation shown in Figure 10 is proposed in this work. In this implementation, there is neither V_{PV} nor I_{PV} sensors and it is assumed that the power injected into the grid is almost equal to the power extracted from the PV panels, $P_G \approx P_{PV}$, as in Reference [11]. A novelty of this research is that it takes de advantages of the PCC of the DC-DC stage and some SOGI based enhancements applied to the control of the VSI stage to improve the performance of the MPPT implementation.

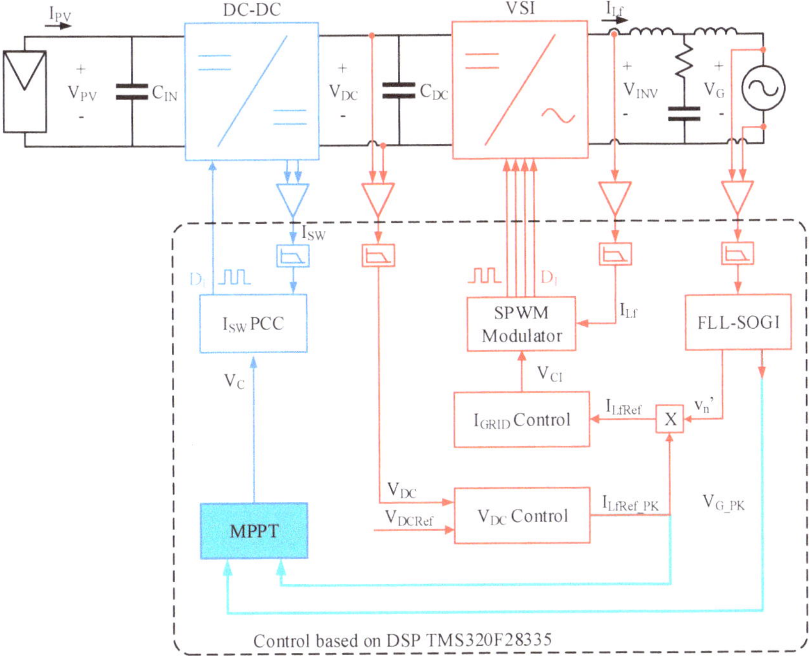

Figure 10. Proposed sensorless MPPT implementation.

A perturb and observe (P and O) algorithm [36] has been programmed in both the conventional (Figure 10) and the proposed sensorless (Figure 11) MPPT algorithms to compare the performance of both implementations. Both implementations use the grid frequency as a time-base to execute the MPPT algorithm. This technique increases the rejection of the disturbances caused by V_{DC_R}. In previous applications of a similar technique [12], the MPPT algorithm was executed at twice the grid frequency, but that sampling frequency is too fast for the control loops implemented in the proposed sensorless algorithm. In the present study, the MPPT algorithm was executed once every five cycles of the grid, yielding $F_{MPPT} = 10$ Hz.

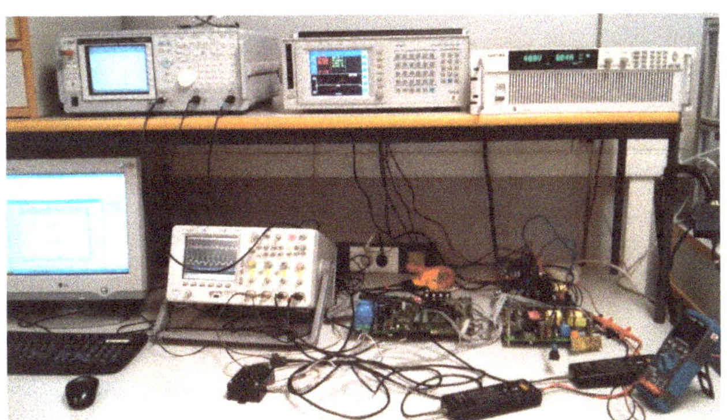

Figure 11. Experimental setup.

In the conventional implementation, the MPPT algorithm provides the reference of V_{PV} (V_{PV_Ref}) to the V_{PV} control loop. In the proposed sensorless MPPT, both the V_{PV} sensor and the V_{PV} control loop have been removed, so that the MPPT provides the reference Vc to the analog PCC.

4.1. Estimation of the Power Injected into the Grid in the Sensorless MPPT

The reference $I_{Lf\_Ref\_PK}$ is the peak value of the current injected into the grid, being proportional to P_G when the amplitude of the grid voltage is a static value. The signal $I_{Lf\_Ref\_PK}$ may have a remarkable low-frequency voltage ripple due to V_{DC_R}, thus producing a high distortion of the grid injected current along with a disturbance in the estimation of P_G. The SOGI based notch filter $F_{NS}(s)$ in series with the regulator $G_{VDC}(s)$ shown in Figure 3 is used to filter out that ripple. The use of the notch filter also enables a high crossover frequency of $T_{VDC}(s)$ without increasing the ripple in $I_{Lf\_Ref\_PK}$, which is useful to implement a fast MPPT algorithm. The crossover frequency of $T_{VDC}(s)$ is $F_{C\_VDC} = 45$ Hz so that an MPPT of $F_{MPPT} = 10$ Hz can be implemented.

The estimation of P_G, P_{G_est}, depends on I_G and on the grid voltage RMS value (V_G) so that variations of V_G perturb the calculations of P_{G_est}. To overcome this issue, the value of the signal v'_{pk} is used to calculate the estimation of P_G as it is shown in Equation (17). The signal v'_{pk} is the amplitude of the fundamental of V_G and is provided by the FLL-SOGI, not requiring additional computational resources. The signal v'_{pk} has very low sensitivity to the distortion of the grid voltage because it is naturally filtered by the FLL-SOGI.

$$P_{G\_est} = \frac{v'_{pk}}{\sqrt{2}} \cdot \frac{I_{LfRef\_PK}}{\sqrt{2}} = \frac{v'_{pk} \cdot I_{LfRef\_PK}}{2} \approx P_{PV} \tag{17}$$

4.2. Implementation of the Perturb and Observe (P&O) Algorithm

The conventional MPPT algorithm uses the measurements of V_{PV} and I_{PV} to set the operation point of the PV source. The algorithm increases or decreases V_{PV_Ref} in perturbation steps of a value ΔV_{PV_Ref} to move the operation point along the I–V curve. In the sensorless implementation shown in Figure 10, instead of the measurements of V_{PV} and I_{PV}, the value of P_{G_est} expressed by Equation (17) was used. In Reference [11] it was proposed to manage the duty cycle of the switches (D_F) of the DC-DC to move the operation point along the I–V curve. In inverters with a reduced DC-link, the high susceptibility of V_{PV} to the ripple V_{DC_R} disturbs the operation point in the PV panel.

In the proposed sensorless MPPT, the value of I_{PV} is indirectly set by means of the reference signal of the PCC loop, Vc. The use of PCC has two functions: reducing the susceptibility of V_{PV} to the ripple V_{DC_R} and protecting the DC-DC converter from overcurrents.

The variable Vc is increased or decreased in small steps of a value ΔVc. It is worth pointing out that an increase of Vc causes an increase of I_{PV}, moving the PV operation point to the left of the I–V curve, whereas a decrease of Vc moves the PV operation point to the right of the I–V curve.

5. Results

Figure 11 depicts the experimental setup. The laboratory tests have been performed using the two-stage inverter presented in Figure 1. The inverter under test was designed for connecting a single PV panel of 230 W to the single phase grid (230 V_{RMS} @ 50 Hz) and has a DC-link with the capacitance calculated in (1), $C_{DC} = 50$ µF. The challenge of using such a small value of C_{DC} is to keep a low distortion of the grid current and small transient overvoltages at the DC-link. The control of the two-stage inverter has been implemented digitally in a Texas Instruments TMS320F28335 DSP [30] at a sampling frequency (F_S) of 40 kHz.

The grid was emulated by means of a Cinergia GE&EL 50 grid emulator and electronic load. The voltage waveform was programmed according to the test waveform described in the international standard IEC-61000-4-7 [37], which has a value: $THD_V = 1.2\%$. The PV panel has been emulated by means of an AMETEK TerraSAS ETS1000X10D PV simulator.

5.1. Control of the VSI

5.1.1. Transients of the DC-link Voltage

Figure 12 shows the transient response of the DC-link voltage (V_{DC}) and the current injected into the grid (I_G) when the PV power steps from 150 W up to 200 W. The results shown in Figure 12a were obtained with a crossover frequency Fc_VDC = 10 Hz, yielding an overvoltage in V_{DC} of 53V from its steady-state value. The response in Figure 12b is obtained with a crossover frequency Fc_VDC = 45 Hz. The response in Figure 12b was close to five times faster and the overvoltage is only 15 V, which represents a reduction of 72%.

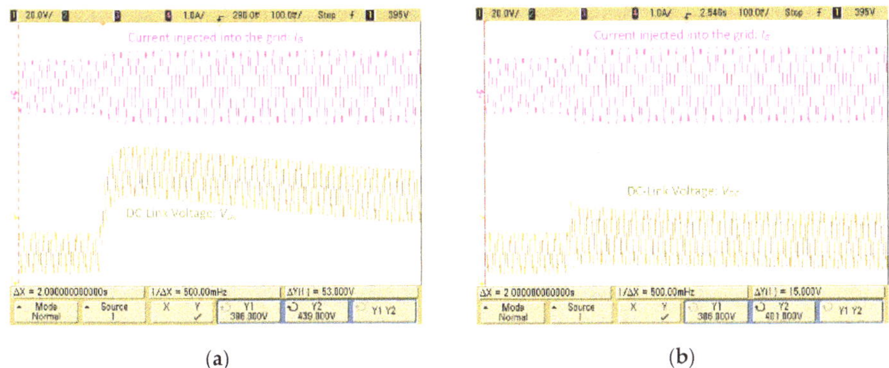

(a) (b)

Figure 12. Power step from 150 W to 200 W. C_{DC}: 50 µF. (**a**): Fc_{TVDC}: 10 Hz. (**b**): Fc_{TVDC}: 45 Hz.

5.1.2. Influence of the SOGI Notch in the Distortion of the Current Injected to the Grid

The increment of the crossover frequency Fc_VDC involves a higher susceptibility to V_{DC_R}, which increases the harmonic distortion of the current injected into the grid (THDi). Figure 13 depicts the current injected into the grid (I_G) using two different regulators for controlling V_{DC}. In Figure 13b, the regulator $G_{VDC}(s)$ of the V_{DC} control loop is the PI shown in Equation (9) with a crossover frequency Fc_VDC = 53 Hz. The waveform of I_G in Figure 13b was obtained using the same PI regulator, but in series with the SOGI notch filter $F_{NS}(s)$ shown in Equation (10), centered at 100 Hz. The crossover frequency has been slightly reduced, Fc_VDC = 45 Hz, taking into account the addition of $F_{NS}(s)$.

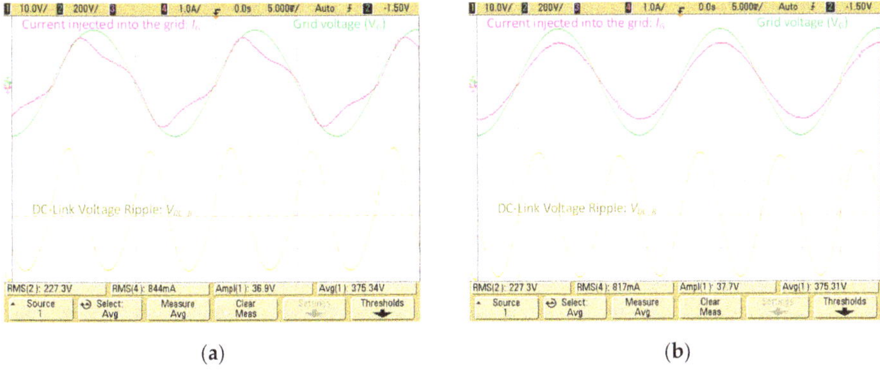

(a) (b)

Figure 13. THDi of the current injected into the grid at P_G = 200 W. Grid: EN-61000-4-7 (1.2% THDv). Fc_VDC = 50 Hz. (**a**): PI Reg.: $G_{VDC}(s)$. (**b**): PI Reg. + Notch filter: $G_{VDC}(s) + F_{NS}(s)$.

Both tests have been performed injecting 200 W to the grid. The controller of the current injected into the grid (I_G) is formed by the P+R regulator in series with the HC expressed by Equation (6). The resulting values of the THDi are 21.51% and 0.96%, respectively. This result indicates the effectiveness of the notch filter for reducing the THDi in spite of the small DC-link.

The following tests of the THDi have been performed for values of P_G in the range P_G = [40, 180] W. The results are shown in Figure 14. The purple trace represents the values of THDi obtained without the SOGI notch $F_{NS}(s)$ and the green trace represents the values obtained with $F_{NS}(s)$. The THDi without notch varies from 21.51% to 22.43%, clearly exceeding the limits of the IEEE519 standard [4] (5%), shown by the red line. The values of the THDi obtained with $F_{NS}(s)$ vary from 0.96% to 3.14%, widely complying with IEEE519 in the whole range of P_G values. In all the measurements the distortion of the grid voltage is THDv = 1.2%.

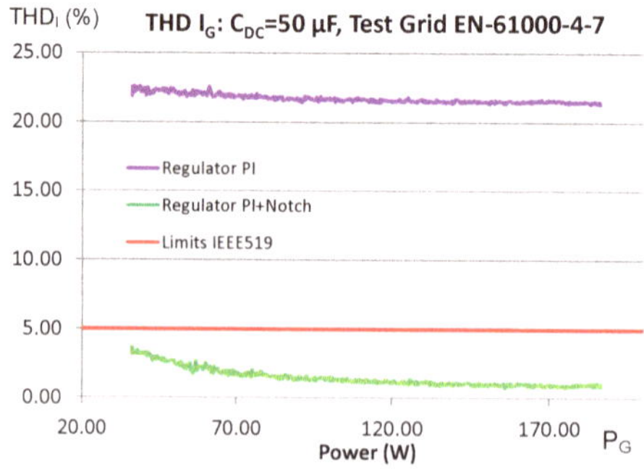

Figure 14. THD of I_G. Grid: EN-61000-4-7 (1.2% THDv). Linear sweep of power injected into the grid (P_G) from 40 W to 180 W at grid frequency F_G = 50 Hz.

5.1.3. Loop Gain Measurement

The loop gain $T_{VDC}(s)$ has been validated by means of loop gain measurement procedures [38–40]. The setup of the test is shown in Figure 15. An NF FRA5097 frequency response analyzer (FRA) is configured to perform an AC sweep from 2 Hz to 20 kHz. The signal generated by the oscillator of the FRA (v_{OSC_A}) is acquired by the DSP, which carries out the control of the inverter through an internal 12-bit ADC. The acquired signal (v_{OSC}) was digitally injected into the control loop as a perturbation. Both, v_{DC} and $v_{DC}+v_{OSC}$ signals were adapted digitally to be loaded into the pulse width modulation (PWM) unit of the DSP. The offset of the signals was removed through digital high pass filters, and then the amplitudes are digitally adjusted to maximize the resolution of the PWM. The PWM signals ($v_{DC})_{PWM}$ and ($v_{DC} + v_{OSC})_{PWM}$ were measured by the FRA. The PWM signals were filtered by the FRA through its internal tracking filter. The loop gain measurement of $T_{VDC}(s)$, shown in Figure 16, was performed at P_G = 230 W. The results show the similarity between the experimental and theoretical Bode plots of $T_{VDC}(s)$.

5.2. MPPT

The performance of the sensorless "perturb and observe" (P and O) MPPT algorithm presented in this study has been compared to a conventional implementation, which uses sensors to measure V_{PV} and I_{PV}. The experimental tests have carried out to measure the start-up time until the maximum power point (MPP) was reached, and the performance at the MPPT under irradiation transients. All the tests were performed with an MPPT sampling frequency, F_{MPPT} = 10 Hz. The perturbation step value programmed in the conventional implementation was ΔV_{PV_Ref} = 300 mV. In the sensorless MPPT the perturbation step was ΔVc = 12.5 mV, which corresponds to a 250 mA step in I_{PV} in the operation region close to the MPP. It is worth pointing out that different values are perturbed in both

MPPTs (either V_{PV_Ref} or ΔV_c) and that the DC-DC converter operates in discontinuous conduction mode. Both facts prevent finding an equivalent value of the perturbation step in both MPPTs.

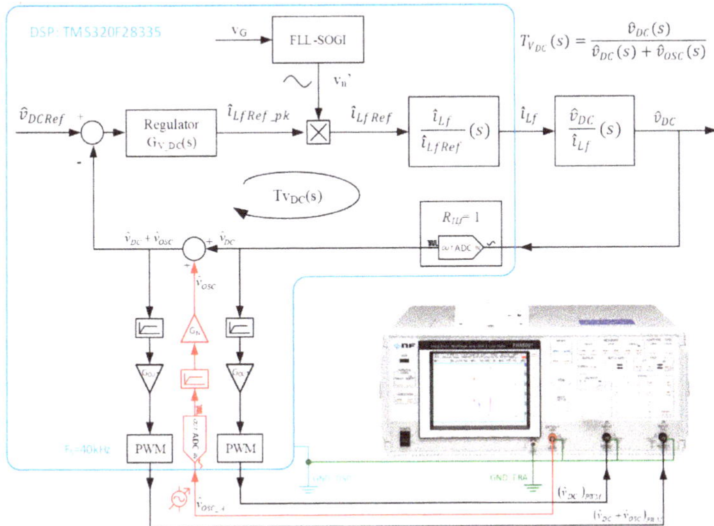

Figure 15. Loop gain measurement setup.

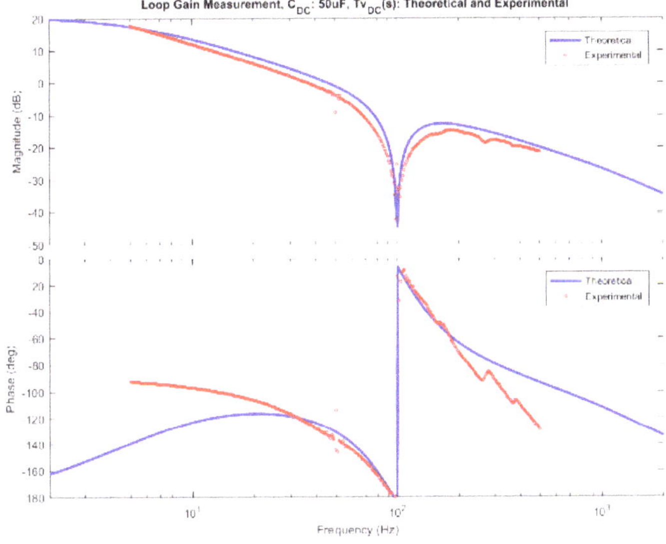

Figure 16. Loop gain frequency response of $Tv_{DC}(s)$. $P_G = 230$ W, $C_{DC} = 50$ µF. Blue: theoretical, red: experimental.

5.2.1. Start-up Time to Reach the MPP

Figure 17 shows the evolution of the operation point at the PV source (V_{PV}, I_{PV} and P_{PV}) from the start-up until the MPP was reached. The conventional implementation was faster (2.75 s) than the sensorless (12.6 s) due to the differences in the perturbation step, but once the MPP is achieved, both implementations continue at the MPP.

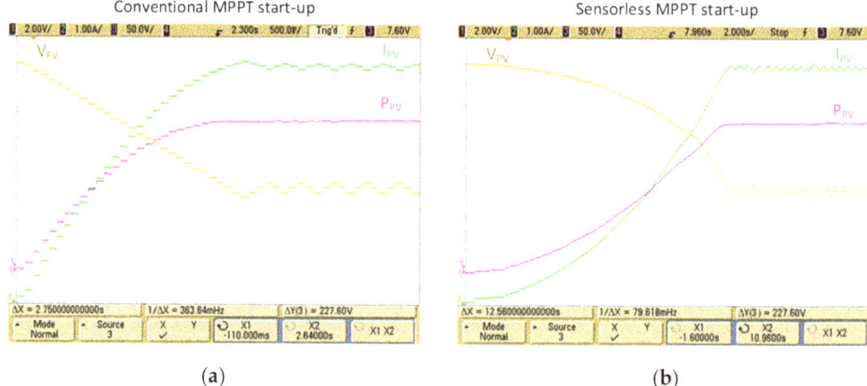

Figure 17. MPPT response from the start-up until the MPP is reached. F_{MPPT}: 10 Hz. (**a**): Conventional, V_{PV} step: ΔV_{PV_Ref} = 300 mV. (**b**): Sensorless, PCC Vc step: ΔVc = 12.5 mV ($\approx$250 mA close to the MPP).

5.2.2. MPPT Performance Close to the MPP

A key factor in the MPPT performance is the accuracy of the PV estimation and the dispersion of the operation point from the MPP along the I–V curve. The results shown in Figure 18 were obtained tracking the MPP of the 230 W PV panel at constant irradiation of 1000 W/m^2 during 50 s. The results in Figure 18a correspond to the conventional MPPT algorithm and the results in Figure 18b correspond to the sensorless algorithm. In both implementations, the operation points are very close to the MPP. The power extracted from the PV panel during the tests is shown in Table 4. In order to calculate the MPP tracking performance, the quotient between the average power and the peak power measured in each experiment has been used as a reference, as it is expressed by Equation (18). Although the sensorless algorithm presents a very slight disadvantage (0.06%) in terms of the tracking efficiency, the performance of both methods is almost the same.

$$\text{Tracking efficiency} = \frac{\text{Average Power}}{\text{Peak Power}}. \tag{18}$$

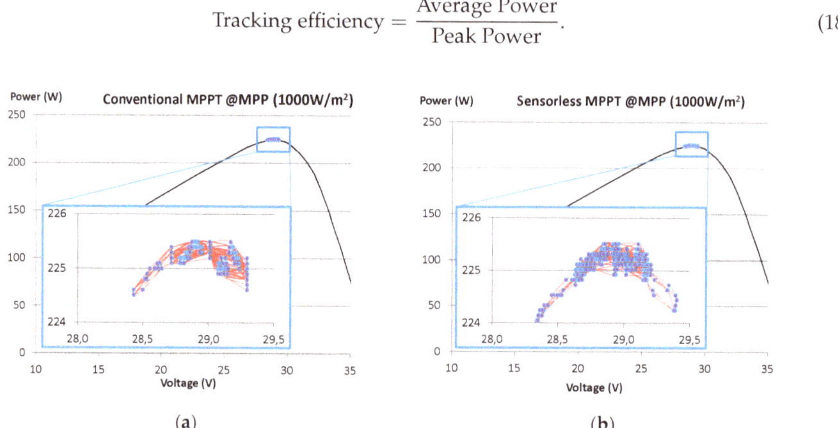

Figure 18. Dispersion of the operation point near the MPP. Total test time: 50 s. F_{MPPT} = 10 Hz. (**a**): Conventional, V_{PV} step: ΔV_{PV_Ref} = 300 mV. (**b**): Sensorless, PCC Vc step: ΔVc = 12.5 mV ($\approx$250 mA near the MPP).

Table 4. Performance of the MPPT algorithms at the MPP under constant irradiation (1000 W/m^2).

MPPT Implementation	Conventional	Sensorless
Average power (W)	225.30	224.80
Peak power (W)	225.49	225.11
Tracking efficiency (%)	99.92	99.86

The tracking of the MPP under heavy irradiation transients is shown in Figure 19. The tests have been performed for 50 s following the stages shown in Table 5. The black traces are the IV curve of the PV panel at 1000 W/m^2 and 600 W/m^2. The blue traces represent the evolution in stage 2, during the reduction of the irradiation, and the red trace represents the evolution in stage 4, during the increase of irradiation.

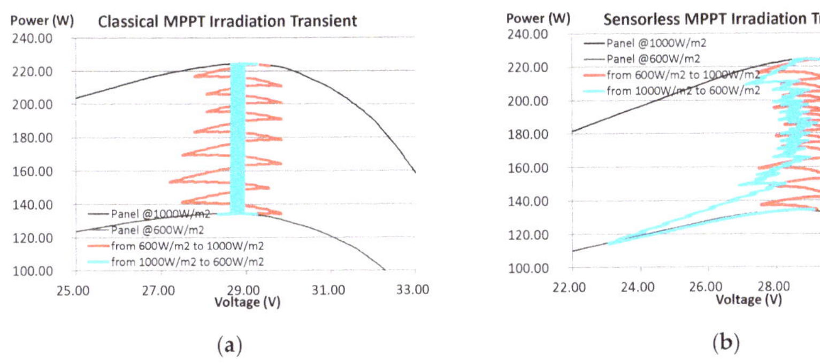

Figure 19. Irradiation transients test. Total time: 50 s. F_{MPPT} = 10 Hz. (**a**) Conventional, V_{PV} step: ΔV_{PV_Re} = 300 mV. (**b**) Sensorless, PCC Vc step: ΔVc = 12.5 mV ($\approx$250 mA near the MPP).

Table 5. Stages of the transient radiation test.

Stage	Irradiation Profile	Duration
1	Constant at 1000 W/m^2	10 s
2	Linear decrease from 1000 W/m^2 to 600 W/m^2	10 s
3	Constant at 600 W/m^2	10 s
4	Linear increase from 600 W/m^2 to 1000 W/m^2	10 s
5	Constant at 1000 W/m^2	10 s
Total	-	50 s

The power extracted from the PV panel during the tests is shown in Table 6. During stages 1, 3 and 5, both MPPTs tracked the MPP as was expected from the results shown in Table 4. During stage 2, the sensorless MPPT was not as accurate as the conventional one, but during stage 4 the sensorless MPPT exhibited a smaller dispersion of the operation point and, thus, a higher accuracy than the conventional one. It can be stated that the MPPT performance of both MPPTs was highly similar. Although the conventional MPPT had a faster start-up and was slightly more accurate in some tests, the performance of the proposed sensorless MPPT near the MPP was almost equal to the conventional implementation, but at a lower cost.

Table 6. Transient radiation test results.

Stage	Conventional MPPT	Sensorless MPPT	%
1	225.30 W	224.80 W	0.998
2	181.27 W	173.51 W	0.957
3	134.90 W	134.41 W	0.996
4	180.82 W	180.36 W	0.997
5	225.30 W	224.80 W	0.998
Average	189.52 W	187.58 W	0.989

6. Conclusions

This paper focused on control techniques which help to reduce the cost of two-stage grid-connected PV inverters. In previous works, sensorless algorithms and techniques to reduce the capacitance of the DC-link have been proposed separately. The present study attempts to integrate both trends in a single implementation. The reduction of the DC-link capacitance requires a faster control loop to keep the DC-link voltage within safe values. However, this practice usually increases the THD of the current injected into the grid. The DC-link can be reduced in a factor of ten compared to standard values of the DC-link capacitance, yet with good values of the THDi, by increasing the speed of the voltage control loop and using a frequency adaptive notch filter tuned at twice the grid frequency.

SOGI structures are an effective way to implement tuned filters both in the inverter voltage loop and in its current loop. It is shown that the crossover frequency of the DC-link voltage control loop can be increased from typical values of 10 Hz to a value around 50 Hz, yet getting a THDi value lower than 1%. The combination of a fast voltage loop with a SOGI notch filter allows the reduction of the DC-link capacitance. An FLL-SOGI is used to get the value of the grid frequency to tune the SOGI controllers in the inverter control loops.

Summing up, this study proposes the implementation of the MPPT with no sensors at the PV source side, which takes advantages of the SOGI based enhancements implemented in the control of the converter. The high dynamics achieved by the inverter controllers yield a performance of the sensorless MPPT very similar to that of conventional MPPT implementations, but at a lower cost.

Author Contributions: R.G.-M., G.G. and E.F. proposed the main idea, performed the investigation and designed the experiments; R.G.-M. and G.G. developed the software, performed the experiments, and wrote the paper. M.L. and S.M. processed the data from the experimental results and reviewed the paper. G.G. and E.F. lead the project, acquired the funds for research.

Funding: This work has been co-financed by the Spanish Ministry of Economy and Competitiveness (MINECO) and by the European Regional Development Fund (ERDF) under Grant ENE2015-64087-C2-2-R and by the Spanish Ministry of Science, Innovation and University (MICINN) and the European Regional Development Fund (ERDF) under Grant RTI2018-100732-B-C21. The European Regional Development Fund (ERDF) and the Generalitat Valenciana (GVA) financed the purchase of the Cinergia GE&EL 50 grid emulator and electronic load used during the experimental part of this work under Grant IDIFEDER/2018/036.

Conflicts of Interest: The authors declare no conflict of interest. The funders had no role in the design of the study; in the collection, analyses, or interpretation of data; in the writing of the manuscript, or in the decision to publish the results.

Appendix A

Table A1 shows the approximated costs of the prototype built for this study at the present time. It is important to note that a 6% cost saving is estimated.

Table A1. Approximate costs of the prototype.

Implementation	Conventional	Sensorless and Reduced DC-link	
PCB	100 €	100 €	100 €
Common components	400 €	400 €	400 €
DC-link	Electrolytic 500 μF 12 €	Electrolytic 50 μF 5 €	Film 50 μF 7 €
V_{PV} and I_{PV} sensors	25 €	0 €	0 €
Total	537 €	505 € (94%)	507 € (94.4%)

References

1. International Electrotechnical Commission. *Electromagnetic Compatibility (EMC)—Part 3-2: Limits—Limits for Harmonic Current Emissions (Equipment Input Current ≤16 A Per Phase)*; International Electrotechnical Commission (IEC): Geneva, Switzerland, 2018.

2. International Electrotechnical Commission. *Electromagnetic Compatibility (EMC)—Part 3-4: Limits—Limitation of Emission of Harmonic Currents in Low-Voltage Power Supply Systems for Equipment with Rated Current Greater than 16 A*; International Electrotechnical Commission (IEC): Geneva, Switzerland, 1998.

3. International Electrotechnical Commission. *Electromagnetic compatibility (EMC)—Part 3-12: Limits—Limits for Harmonic Currents Produced by Equipment Connected to Public Low-Voltage Systems with Input Current >16 A and ≤75 A Per Phase*; International Electrotechnical Commission (IEC): Geneva, Switzerland, 2011.

4. Langella, R.; Testa, A.; Alii, E. IEEE Std 519-2014—IEEE Recommended Practice and Requirements for Harmonic Control in Electric Power Systems. *IEEE Trans. Ind. Appl.* **1989**, *25*, 6–7.

5. Espi, J.M.; Castello, J. A Novel Fast MPPT Strategy for High Efficiency PV Battery Chargers. *Energies* **2019**, *12*, 1152. [CrossRef]

6. Ahmed, J.; Salam, Z. A Modified P&O Maximum Power Point Tracking Method With Reduced Steady-State Oscillation and Improved Tracking Efficiency. *IEEE Trans. Sustain. Energy* **2016**, *7*, 1506–1515.

7. Sher, H.A.; Rizvi, A.A.; Addoweesh, K.E.; Al-Haddad, K. A Single-Stage Stand-Alone Photovoltaic Energy System With High Tracking Efficiency. *IEEE Trans. Sustain. Energy* **2017**, *8*, 755–762. [CrossRef]

8. Liivik, E.; Chub, A.; Kosenko, R.; Vinnikov, D. Low-cost photovoltaic microinverter with ultra-wide MPPT voltage range. In Proceedings of the 2017 6th International Conference on Clean Electrical Power (ICCEP), Santa Margherita Ligure, Italy, 27–29 June 2017; pp. 46–52. [CrossRef]

9. Sher, H.A.; Addoweesh, K.E.; Haddad, K.A. An Efficient and Cost-Effective Hybrid MPPT Method for a Photovoltaic Flyback Micro-Inverter. *IEEE Trans. Sustain. Energy* **2017**, *9*, 1137–1144. [CrossRef]

10. Samrat, P.S.; Edwin, F.F.; Xiao, W. Review of current sensorless maximum power point tracking technologies for photovoltaic power systems. In Proceedings of the 2013 International Conference on Renewable Energy Research and Applications (ICRERA), Madrid, Spain, 20–23 October 2013; pp. 20–23. [CrossRef]

11. Kitano, T.; Matsui, M.; Xu, D.-H. Power sensor-less MPPT control scheme utilizing power balance at DC link-system design to ensure stability and response. In Proceedings of the IECON 01—27th Annual Conference of the IEEE Industrial Electronics Society, Denver, CO, USA, 29 November–2 December 2001; Volume 2, pp. 1309–1314.

12. Choi, B.-Y.; Jang, J.-W.; Kim, Y.-H.; Ji, Y.-H.; Jung, Y.-C.; Won, C.-Y. Current sensorless MPPT using photovoltaic AC module-type flyback inverter. In Proceedings of the 2013 IEEE International Symposium on Industrial Electronics (ISIE), Taipei, Taiwan, 28–31 May 2013; pp. 28–31. [CrossRef]

13. Metry, M.; Shadmand, M.B.; Balog, R.S.; Abu-Rub, H. MPPT of Photovoltaic Systems Using Sensorless Current-Based Model Predictive Control. *IEEE Trans. Ind. Appl.* **2017**, *53*, 1157–1167. [CrossRef]

14. Femia, N.; Petrone, G.; Spagnuolo, G.; Vitelli, M. A Technique for Improving P&O MPPT Performances of Double-Stage Grid-Connected Photovoltaic Systems. *IEEE Trans. Ind. Electron.* **2009**, *56*, 4473–4482. [CrossRef]

15. Garcerá, G.; González-Medina, R.; Figueres, E.; Sandia, J. Dynamic modeling of DC–DC converters with peak current control in double-stage photovoltaic grid-connected inverters. *Int.J. Circuit Theory Appl.* **2012**, *40*, 793–813. [CrossRef]

16. Zhu, G.; Ruan, X.; Zhang, L.; Wang, X. On the Reduction of Second Harmonic Current and Improvement of Dynamic Response for Two-Stage Single-Phase Inverter. *IEEE Trans. Power Electron.* **2015**, *30*, 1028–1041. [CrossRef]

17. Karanayil, B.; Agelidis, V.G.; Pou, J. Performance Evaluation of Three-Phase Grid-Connected Photovoltaic Inverters Using Electrolytic or Polypropylene Film Capacitors. *IEEE Trans. Sustain. Energy* **2014**, *5*, 1297–1306. [CrossRef]

18. Schonberger, J. A single phase multi-string PV inverter with minimal bus capacitance. In Proceedings of the 2009 13th European Conference on Power Electronics and Applications, Barcelona, Spain, 8–10 September 2009; pp. 1–10. [CrossRef]

19. Messo, T.; Jokipii, J.; Suntio, T. Minimum DC-link capacitance requirement of a two-stage photovoltaic inverter. In Proceedings of the 2013 IEEE Energy Conversion Congress and Exposition, Denver, CO, USA, 15–19 September 2013; pp. 999–1006. [CrossRef]

20. Heo, J.; Matsuo, K.; Hen, P.; Kondo, K. Dynamics of a minimum DC link voltage driving method to reduce system loss for hybrid electric vehicles. In Proceedings of the 2017 IEEE International Electric Machines and Drives Conference (IEMDC), Miami, FL, USA, 21–24 May 2017; pp. 1–7.

21. Karanayil, B.; Agelidis, V.G.; Pou, J. Evaluation of DC-link decoupling using electrolytic or polypropylene film capacitors in three-phase grid-connected photovoltaic inverters. In Proceedings of the IECON 2013—39th Annual Conference of the IEEE Industrial Electronics Society, Vienna, Austria, 10–13 November 2013; pp. 6980–6986. [CrossRef]

22. Bazargan, D.; Bahrani, B.; Filizadeh, S. Reduced Capacitance Battery Storage DC-link Voltage Regulation and Dynamic Improvement Using a Feedforward Control Strategy. *IEEE Trans. Energy Convers.* **2018**, *33*, 1659–1668. [CrossRef]

23. Ciobotaru, M.; Teodorescu, R.; Blaabjerg, F. A New Single-Phase PLL Structure Based on Second Order Generalized Integrator. In Proceedings of the 2006 37th IEEE Power Electronics Specialists Conference, Jeju, Korea, 18–22 June 2006; pp. 1–6. [CrossRef]

24. Rodriguez, P.; Luna, A.; Ciobotaru, M.; Teodorescu, R.; Blaabjerg, F. Advanced Grid Synchronization System for Power Converters under Unbalanced and Distorted Operating Conditions. In Proceedings of the IECON 2006-32nd Annual Conference on IEEE Industrial Electronics, Paris, France, 7–10 November 2006; pp. 5173–5178. [CrossRef]

25. Rodriguez, P.; Luna, A.; Candela, I.; Teodorescu, R.; Blaabjerg, F. Grid synchronization of power converters using multiple second order generalized integrators. In Proceedings of the 2008 34th Annual Conference of IEEE Industrial Electronics, Orlando, FL, USA, 10–13 November 2008; pp. 755–760. [CrossRef]

26. Rodriguez, P.; Luna, A.; Etxeberria, I.; Hermoso, J.R.; Teodorescu, R. Multiple second order generalized integrators for harmonic synchronization of power converters. In Proceedings of the 2009 IEEE Energy Conversion Congress and Exposition, 2009 ECCE, San Jose, CA, USA, 20–24 September 2009; pp. 2239–2246. [CrossRef]

27. Rodríguez, P.; Luna, A.; Candela, I.; Mujal, R.; Teodorescu, R.; Blaabjerg, F. Multiresonant Frequency-Locked Loop for Grid Synchronization of Power Converters Under Distorted Grid Conditions. *IEEE Trans. Ind. Electron.* **2011**, *58*, 127–138. [CrossRef]

28. Liserre, M.; Blaabjerg, F.; Hansen, S. Design and control of an LCL-filter-based three-phase active rectifier. *IEEE Trans. Ind. Electron.* **2009**, *56*, 4492–4501. [CrossRef]

29. TMS320F28335 Delfino™ 32-bit MCU with 150 MIPS, FPU, 512 KB Flash, EMIF, 12b ADC. Available online: http://www.ti.com/product/TMS320F28335 (accessed on 1 May 2019).

30. Castilla, M.; Miret, J.; Matas, J.; Garcia de Vicuna, L.; Guerrero, J.M. Control Design Guidelines for Single-Phase Grid-Connected Photovoltaic Inverters With Damped Resonant Harmonic Compensators. *IEEE Trans. Ind. Electron.* **2009**, *56*, 4492–4501. [CrossRef]

31. Ciobotaru, M.; Teodorescu, R.; Blaabjerg, F. Control of single-stage single-phase PV inverter. In Proceedings of the 2005 European Conference on Power Electronics and Applications, Dresden, Germany, 11–14 September 2005; pp. 20–26. [CrossRef]

32. Teodorescu, R.; Blaabjerg, F.; Borup, U.; Liserre, M. A new control structure for grid-connected LCL PV inverters with zero steady-state error and selective harmonic compensation. In Proceedings of the Nineteenth Annual IEEE Applied Power Electronics Conference and Exposition, Anaheim, CA, USA, 22–26 February 2004; Volume 1, pp. 580–586. [CrossRef]

33. Reza, M.S.; Ciobotaru, M.; Agelidis, V.G. Grid voltage offset and harmonics rejection using second order generalized integrator and kalman filter technique. In Proceedings of the 2012 7th International Power Electronics and Motion Control Conference (IPEMC), Harbin, China, 2–5 June 2012; pp. 104–111. [CrossRef]
34. Timbus, A.V.; Ciobotaru, M.; Teodorescu, R.; Blaabjerg, F. Adaptive resonant controller for grid-connected converters in distributed power generation systems. In Proceedings of the Twenty-First Annual IEEE Applied Power Electronics Conference and Exposition, Dallas, TX, USA, 19–23 March 2006. [CrossRef]
35. Teodorescu, R.; Liserre, M.; Rodriguez, P. Chapter 9: Grid Converter Control for WTS. In *Grid Converters for Photovoltaic and Wind Power Systems*; Wiley-IEEE Press: Chichester, England, 2010; pp. 205–235. ISBN 978-0-470-05751-3.
36. Hohm, D.P.; Ropp, M.E. Comparative study of maximum power point tracking algorithms. *Prog. Photovolt. Res. Appl.* **2003**, *11*, 47–62. [CrossRef]
37. International Standard IEC 61000-4-7. *Electromagnetic Compatibility (EMC)—Part 4-7:Testing and Measurement Techniques—General Guide on Harmonics and Interharmonics Measurements and Instrumentation, for Power Supply Systems and Equipment Connected Thereto*; International Electrotechnical Commission (IEC): Geneva, Switzerland, 2002.
38. Middlebrook, R.D. Measurement of Loop Gain in Feedback Systems. *Int. J. Electron.* **1975**, *38*, 485–512. [CrossRef]
39. Gonzalez-Espin, F.; Figueres, E.; Garcera, G.; Gonzalez-Medina, R.; Pascual, M. Measurement of the Loop Gain Frequency Response of Digitally Controlled Power Converters. *IEEE Trans. Ind. Electron.* **2010**, *57*, 2785–2796. [CrossRef]
40. Gonzalez-Espin, F.; Figueres, E.; Garcera, G.; Gonzalez, R.; Carranza, O.; Gonzalez, L.G. A digital technique to measure the loop gain of power converters. In Proceedings of the 13th European Conference on Power Electronics and Applications, Barcelona, Spain, 8–10 September 2009; pp. 1–10.

Article

A Novel Fast MPPT Strategy for High Efficiency PV Battery Chargers

Jose Miguel Espi * and Jaime Castello

Department of Electrical Engineering, University of Valencia, Avd. Universitat S/N,
46100 Burjassot-Valencia, Spain; jaime.castello@uv.es
* Correspondence: jose.m.espi@uv.es; Tel.: +34-963-543-450

Received: 26 February 2019; Accepted: 19 March 2019; Published: 25 March 2019

Abstract: The paper presents a new maximum power point tracking (MPPT) method for photovoltaic (PV) battery chargers. It consists of adding a low frequency modulation to the duty-cycle and then multiplying the ac components of the panel voltage and power. The obtained parameter, proportional to the conductance error, is used as a gain for the integral action in the charging current control. The resulting maximum power point (MPP) is very still, since the integral gain tends to zero at the MPP, yielding PV efficiencies above 99%. Nevertheless, when the operating point is not the MPP, the integral gain is large enough to provide a fast convergence to the MPP. Furthermore, a fast power regulation on the right side of the MPP is achieved in case the demanded power is lower than the available maximum PV power. In addition, the MPPT is compatible with the control of a parallel arrangement of converters by means of a droop law. The MPPT algorithm gives an averaged duty-cycle, and the droop compensation allows duty-cycles to be distributed to all active converters to control their currents individually. Moreover, the droop strategy allows activation and deactivation of converters without affecting the MPP and battery charging operation. The proposed control has been assayed in a battery charger formed by three step-down converters in parallel using synchronous rectification, and is solved in a microcontroller at a sampling frequency of 4 kHz. Experimental results show that, in the worst case, the MPPT converges in 50 ms against irradiance changes and in 100 ms in case of power reference changes.

Keywords: photovoltaic (PV); maximum power point (MPP); maximum power point tracking (MPPT), perturbe and observe (P&O); incremental conductance (IC)

---

1. Introduction

Photovoltaic (PV) battery chargers are designed to maximize the energy extracted from solar panels. This requires the maximization of both electronic and PV efficiencies. The electronic efficiency is increased by a parallel configuration of multiple power converters and a synchronous rectification implemented in each one. Paralleling permits activating/deactivating the converters so that each active converter works near its nominal power, thus saving the conduction and switching losses of the inactive converters. Moreover parallelized solutions allows power scaling and increases reliability. On the other hand, the PV efficiency is defined as the ratio between the average power extracted from the panels and the maximum power that can be extracted at a given irradiance. Maximum power point tracking (MPPT) algorithms automatically adjust the PV voltage at the converter input to get the maximum power for each present irradiance level. When a change in irradiance occurs, an ideal MPPT algorithm should reach the new maximum power point (MPP) as fast as possible, and then remain at the MPP without fluctuations. However, in practice, the MPPT algorithms exhibit oscillations around the MPP and take a certain time to converge, penalizing the PV efficiency.

The MPPT strategies can be classified into two main categories: the stand-alone MPPTs [1–5] and the converter-embedded MPPTs [6–10]. A stand-alone MPPT is an independent module that uses the

PV voltage and current to determine the input voltage reference to be transmitted to all converters installed. This is typically implemented using perturb and observe (P&O) algorithms. The advantage of the stand-alone method is that it can be used to manage parallelized converters without having to modify their respective controls. In contrast, a converter-embedded MPPT is programmed in the converter control to determine directly the duty-cycle that maximizes the PV power. It is usually implemented using incremental conductance (IC) algorithms [6–8]. Converter-embedded strategies are much faster than stand-alone strategies, thus presenting a higher PV dynamic efficiency, which makes them more suitable in applications where irradiance changes are fast and frequent [11]. However, the parallel multi-stage arrangement becomes difficult to control using a converter-embedded MPPT, as it calculates a single duty-cycle that maximizes the PV power.

In recent research [6,7], new converter-embedded MPPT strategies based on IC have been presented for a step-up converter that combines a fast convergence with a small fluctuation around the MPP. In [7], the static g_{dc} and dynamic g_{ac} PV conductances are explicitly calculated using a moving average filter (MAF) and a lock-in amplifier (LIA) respectively, and then compared and regulated to be equal using an integrator. The ac components used to calculate g_{ac} are the switching ripple components of the PV voltage and current. The MPPT converges to the MPP in approximately 400 ms. As the method requires a small input capacitance to measure g_{ac}, the current ripple is present in the PV current and the MPP fluctuates at the switching frequency. To minimize this effect, a large inductance was utilized to achieve a PV efficiency of 99%. More recently, in [6], the IC is solved in a traditional way by incrementing or decrementing the duty-cycle depending on the sign of the conductance error $g_{ac} - g_{dc}$. The ac components used to evaluate the incremental PV magnitudes are the natural oscillations of the input filter. The MPPT settling time was around 300 ms, and the MPP oscillates at the natural frequency of the filter, resulting in a PV efficiency of 97.5%. In both papers, the high frequency used to calculate the PV AC components makes high frequency sampling rates of above 100 kHz necessary, which increases hardware cost and complexity.

This paper presents a new MPPT strategy for step-down battery chargers that combines the benefits of both stand-alone and converter-embedded methods. The proposed MPPT is integrated with the proportional-integral (PI) current regulator, offering a fast convergence to the MPP in less than 100 ms when the demanded power is higher than available maximum PV power, with smooth transitions to regulation when the required power is lower than the available PV power. The basic idea is to insert the conductance error as an additional gain for the PI's integrator when tracking the MPP. This leads to a fast MPPT but a still MPP with PV efficiency higher than 99%, since the conductance error is null at the MPP. A low frequency modulation of 40 Hz is added to the duty-cycle to get the conductance error by multiplying the AC components of PV power and voltage. As a consequence of the low frequency modulation, the proposed MPPT can be solved at 4 kHz sampling rate by a low-cost microcontroller. Additionally, a droop law is proposed to solve the multiple control of parallel converters, proving that converters' currents can be controlled individually without affecting the MPP operation or battery charge. The proposed MPPT with droop has been assayed in the battery charger shown in Figure 1, where electronic efficiency was improved by means of active rectification using $Q_{R_{j1}}$ and $Q_{R_{j2}}$ transistors and blocking transistors Q_B, instead of using Schottky diodes.

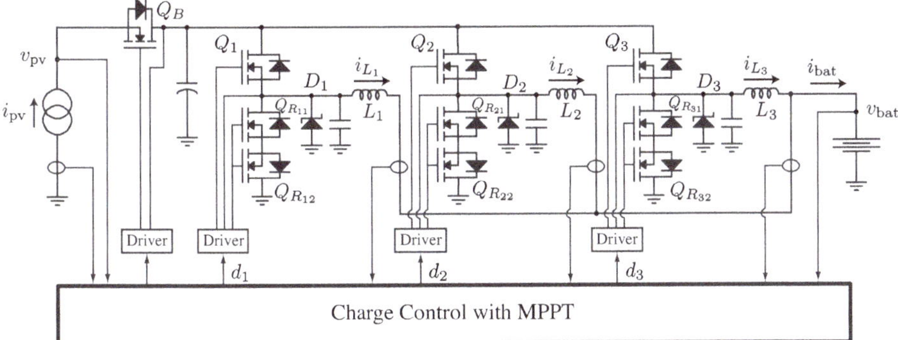

Figure 1. Photovoltaic (PV) battery charger using three parallelized step-down converters with synchronous rectification.

2. Small-Signal Modeling

Figure 2 shows a single step-down converter with synchronous rectification, thus operating always in continuous conduction mode. The circuit shows averaged values of transistors currents, $d \cdot i_L$ and $(1 - d) \cdot i_L$, where d is the converter duty-cycle and r stands for the inductor series resistance. All relationships can be gathered into the block diagram shown in Figure 3, which constitutes a large-signal model. The function f_{pv} solves the current i_{pv} of the PV panel, using the characteristic I-V curve for a given irradiance and voltage v_{pv}. In case of a parallelized step-down converters, d and i_L are vectors containing all duty-cycles and inductor currents, $d \cdot i_L$ is a dot product and $d \cdot v_{\mathrm{pv}}$ is a vector. Notice that, if all converters use the same filtering inductance and receive approximately the same duty-cycle, the presented model is valid just considering that $i_L = \sum i_{L_j} \equiv i_{\mathrm{bat}}$ is the battery charging current, and $L = L_j/n$, $r = r_j/n$, where n is the number of active parallelized converters, since all active inductors can be considered as operating in parallel.

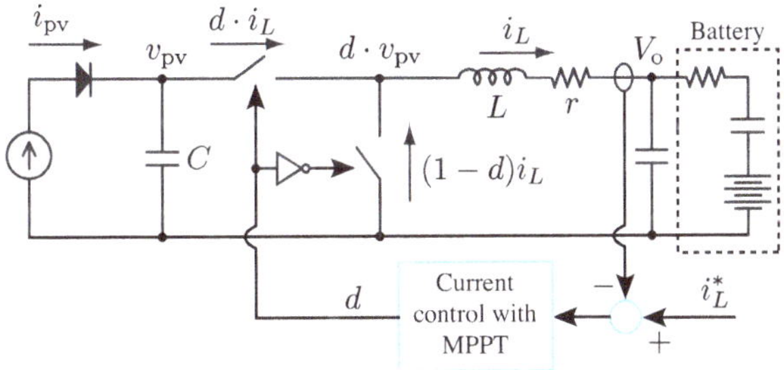

Figure 2. Large-signal averaged circuit of a PV step-down converter with active rectification.

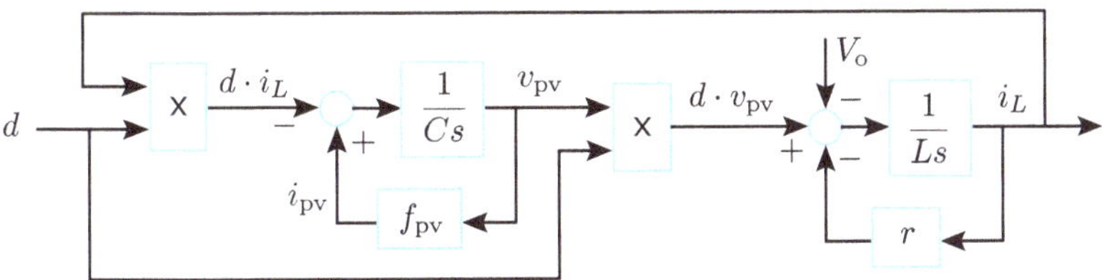

Figure 3. Block diagram of the large-signal model.

As the battery voltage V_o changes much slower than all other circuit variables, it can be assumed constant and the small-signal block diagram results as depicted in Figure 4, where the PV panel voltage and current are related through the incremental conductance $g_{ac} = -\frac{di_{pv}}{dv_{pv}}$. From this figure, the duty-to-voltage and duty-to-current small-signal transfer functions are deduced

$$G_v(s) \equiv \frac{\tilde{v}_{pv}}{\tilde{d}}(s) = \frac{G_{cap}(s) \cdot (I_L + DV_{pv}G_{ind}(s))}{1 - D^2 G_{cap}(s)G_{ind}(s)}, \tag{1}$$

$$G_i(s) \equiv \frac{\tilde{i}_L}{\tilde{d}}(s) = \frac{G_{ind}(s) \cdot (V_{pv} + DI_L G_{cap}(s))}{1 - D^2 G_{cap}(s)G_{ind}(s)}, \tag{2}$$

where $G_{cap}(s) = -1/(Cs + g_{ac})$ and $G_{ind}(s) = 1/(Ls + r)$.

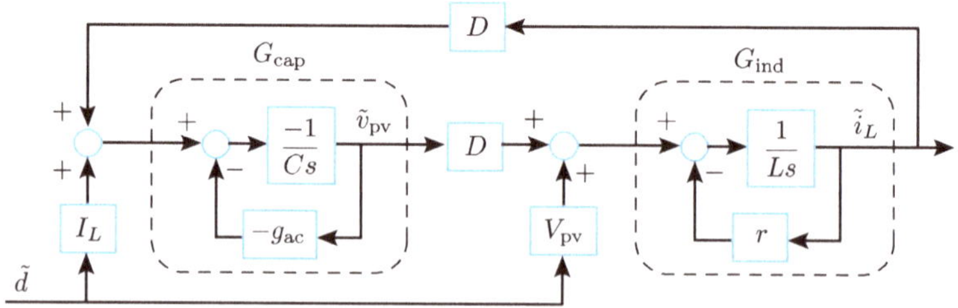

Figure 4. Small-signal model of the power converter and PV panel.

Using the steady-state relationships $DV_{pv} = V_o$ and $DI_L = I_{pv}$, basic manipulations reveal that the small-signal transfer functions (1) and (2) can be expressed as

$$G_v(s) = \frac{-k_v(\frac{s}{\omega_{zv}} + 1)}{(\frac{s}{\omega_n})^2 + 2\zeta(\frac{s}{\omega_n}) + 1}, \tag{3}$$

$$G_i(s) = \frac{k_i(\frac{s}{\omega_{zi}} + 1)}{(\frac{s}{\omega_n})^2 + 2\zeta(\frac{s}{\omega_n}) + 1}, \tag{4}$$

where the natural frequency ω_n and damping factor ζ are

$$\omega_n = \sqrt{\frac{g_{ac}r + D^2}{LC}}, \tag{5}$$

$$\zeta = \frac{1}{2} \cdot \frac{(r/Z_b + g_{ac}Z_b)}{\sqrt{g_{ac}r + D^2}}, \tag{6}$$

being $Z_b = \sqrt{L/C}$. The frequencies of the zeros ω_{zv} and ω_{zi} are

$$\omega_{zv} = \frac{r + DV_o/I_{pv}}{L} \approx \frac{DV_o}{LI_{pv}} = \frac{V_o^2}{P_{pv}L}, \tag{7}$$

$$\omega_{zi} = \frac{g_{ac} - g_{dc}}{C}, \tag{8}$$

where $g_{dc} = \frac{I_{pv}}{V_{pv}}$ is the static conductance, and the dc-gains are

$$k_v = \frac{I_{pv}r/D + V_o}{g_{ac}r + D^2} \approx \frac{V_o}{g_{ac}r + D^2}, \tag{9}$$

$$k_i = \frac{V_{\mathrm{pv}}(g_{\mathrm{ac}} - g_{\mathrm{dc}})}{g_{\mathrm{ac}}r + D^2}. \tag{10}$$

Equation (8) indicates that G_i presents a non-minimum phase zero when operating at the left of the MPP, where $g_{\mathrm{ac}} < g_{\mathrm{dc}}$. In addition, Equation (10) shows that G_i presents null dc-gain when operating exactly at the MPP.

3. Working Principles of the Proposed MPPT

This paper proposes to embed an MPPT strategy in a PI current controller, so that it behaves as a normal PI when the power demanded by $i_{\mathrm{bat}}^* \equiv i_L^*$ is smaller than the present PV power $p_{\mathrm{pv}} = i_{\mathrm{pv}}v_{\mathrm{pv}}$, that is, when $i_{\mathrm{bat}}^* < i_L$, but it opens the current regulation loop and starts MPP tracking when $i_{\mathrm{bat}}^* > i_L$.

Regarding the MPPT strategy, the basic idea is to detect the slope of the P–V curve and use it as a gain for the integral action of the PI. This slope is detected by adding a small amplitude modulation d_m to the duty-cycle

$$d_m(t) = d_{m_{\mathrm{pk}}} \cdot \cos(\omega_m t), \tag{11}$$

where $\omega_m = 2\pi \cdot 40$ rad/s and $d_{m_{\mathrm{pk}}} = 5 \cdot 10^{-3} \equiv 0.5\%$ have been used. According to the duty-to-voltage transfer function G_v, this produces a voltage modulation v_m in the PV panel. As the modulation frequency ω_m is much smaller than ω_{z_v}, it holds that $v_m = -k_v \cdot d_m$. The voltage modulation in turn generates a power modulation p_m as shown in Figure 5.

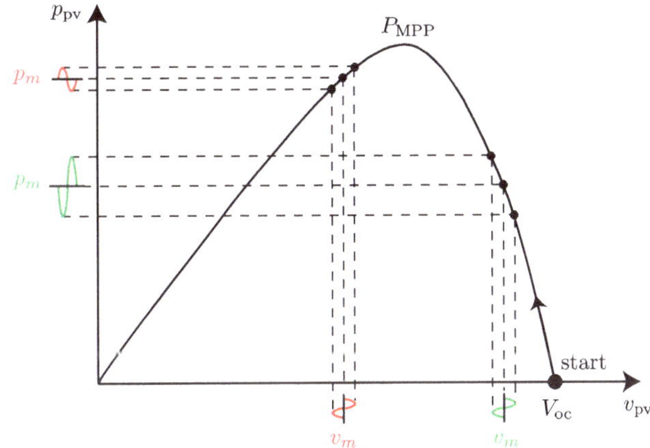

Figure 5. Detection of the PV panel operation: to the left of the MPP (red, with p_m and v_m in-phase) and to the right of the MPP (green, with p_m and v_m in anti-phase).

At a given operating point determined by the PV voltage and current levels $(V_{\mathrm{pv}}, I_{\mathrm{pv}})$, the differential increment of the power is

$$dp_{\mathrm{pv}} = I_{\mathrm{pv}} \cdot dv_{\mathrm{pv}} + V_{\mathrm{pv}} \cdot di_{\mathrm{pv}} \tag{12}$$

and therefore the power slope is

$$\frac{dp_{\mathrm{pv}}}{dv_{\mathrm{pv}}} = I_{\mathrm{pv}} + V_{\mathrm{pv}}\frac{di_{\mathrm{pv}}}{dv_{\mathrm{pv}}} = V_{\mathrm{pv}}(g_{\mathrm{dc}} - g_{\mathrm{ac}}), \tag{13}$$

which reveals the well known incremental conductance condition $g_{\mathrm{dc}} = g_{\mathrm{ac}}$ for any local MPP.

In order for the MPPT algorithm to get the value of this slope, a parameter δ is calculated as

$$\delta(t) \equiv -k_m \cdot p_m(t) \cdot v_m(t). \tag{14}$$

Taking into account that $p_m(t) \approx (\frac{dp_{pv}}{dv_{pv}}) \cdot v_m(t)$ and Equation (13), we get

$$\delta(t) = k_m V_{pv} k_v^2 (g_{ac} - g_{dc}) \cdot d_m^2(t) \tag{15}$$

and, using Equation (11),

$$\delta(t) = \frac{1}{2} k_m V_{pv} (k_v d_{m_{pk}})^2 (g_{ac} - g_{dc})(1 + \cos(2\omega_m t)), \tag{16}$$

which can be separated as $\delta(t) = \delta_{dc} + \delta_{ac}(t)$, where $\delta_{ac} = \delta_{dc} \cdot \cos(2\omega_m t)$ and

$$\delta_{dc} = \frac{1}{2} k_m V_{pv} (k_v d_{m_{pk}})^2 (g_{ac} - g_{dc}). \tag{17}$$

The strategy of the proposed MPPT method is to insert δ as a multiplying factor between the current error e and the PI controller, as shown in Figure 6. When the current error becomes positive, the MPPT is activated by adding the duty-cycle modulation d_m, and the error is multiplied by δ. Only the dc value of δ (17) generates a dc value for the PI to increase or to decrease the duty-cycle. On the left side of the MPP (point 1 in Figure 7a), it holds that $\delta_{dc} < 0$ and therefore the duty-cycle d decreases and v_{pv} increases (Figure 6b). On the right side of the MPP (point 2 in Figure 7a), $\delta_{dc} > 0$ and v_{pv} decrease (Figure 6c). Hence, when the MPPT is activated, the operating point climbs power automatically to a local MPP (yellow point in Figure 7a). The speed of convergence to the MPP is determined by the integrator gain k_{i_l} and the maximum values of δ and e. In the proposed implementation, δ is constrained to the interval $[-1, 1]$ and the error is upper-limited to 1A.

Since $|\delta| < 2|\delta_{dc}|$, and δ_{dc} tends to zero as the operating point approaches the MPP, the power at the MPP is quiescent without oscillations, even if the integrator is set for a fast MPPT, resulting in an excellent static efficiency.

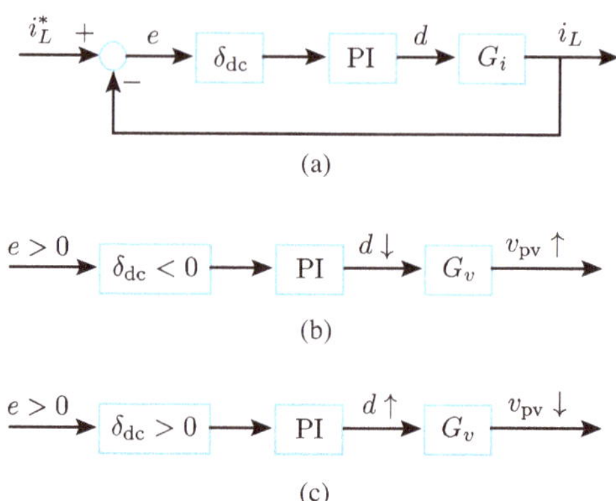

Figure 6. Control block diagram (**a**) and MPP tracking situations: (**b**) to the left of the MPP; and (**c**) to the right of the MPP.

When the current error becomes negative due to an increase in irradiance or a decrease in battery power demand (points 3 or 4 in Figure 7b), d_m is set to zero and δ is set to 1, so the MPPT is transformed into a normal PI current regulation (Figure 6a with $\delta_{dc} = 1$). In this situation, the only possible stable point is on the right side of the MPP (yellow point in Figure 7b).

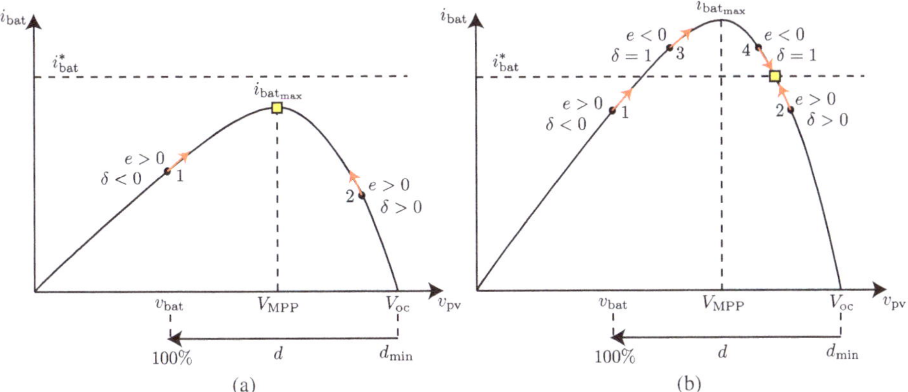

Figure 7. Control performance against the two possible scenarios: (**a**) requested power is higher than available PV power; (**b**) requested power is lower than available PV power.

4. Description of the Implemented Solution

The proposed strategy has been carried out as depicted in Figures 8–11.

Figure 8 shows the typical PI-based control to regulate the battery voltage. The voltage reference v_{bat}^{*} is around 14.7 V/battery during the absorption charge stage, and it is around 13.6 V/battery temperature-compensated float voltage during the float charge stage. The PI determines the charging current i_{bat}, which is limited to i_{max} = 60 A. If the battery SOC is below 80%, this results in a constant current charge at i_{max} and battery voltage below v_{bat}^{*}, while at a higher SOC the battery is charged at a constant voltage v_{bat}^{*} with current below i_{max}. The voltage reference is changed to the temperature-compensated float voltage when charging current is below $10^{-3}C_{10}$ or absorption time exceeds the 8-hour limit.

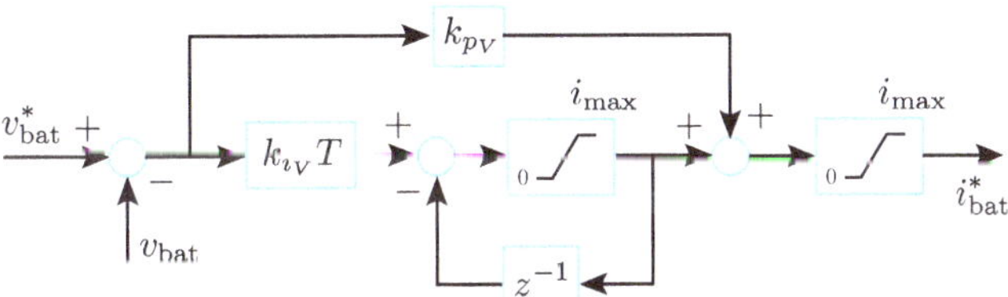

Figure 8. Block diagram of the battery voltage regulation to calculate the battery current reference.

Figure 9 illustrates the implemented strategy to detect the converter working point position relative to the MPP. If the current error is positive and the PV current is higher than i_{start} = 50 mA, the MPPT is started by setting MPPT_ON = 1, and δ is calculated as shown in Equation (14). The voltage v_m and power p_m modulations are extracted from the measured v_{pv} and p_{pv}, respectively, by means of band-pass digital filters $G_{BP}(z)$. These filters were implemented using second-order all-pass filters $G_{AP}(z)$ as

$$G_{BP}(z) = \frac{1}{2}\left[1 - G_{AP}(z)\right] \tag{18}$$

and

$$G_{AP}(z) = \frac{k_2 z^2 + k_1(1 + k_2)z + 1}{z^2 + k_1(1 + k_2)z + k_2}, \tag{19}$$

whose coefficients are calculated as

$$k_1 = -\cos(\omega_0 T),$$
$$k_2 = \frac{1 - \tan(\omega_{\mathrm{BW}} T/2)}{1 + \tan(\omega_{\mathrm{BW}} T/2)}, \tag{20}$$

for a given center-frequency ω_0, bandwidth frequency ω_{BW} and sampling period $T = 1/f_s$. Using $\omega_0 = \omega_m$, $\omega_{\mathrm{BW}} = 2\pi \cdot 80$ rad/s and $f_s = 4$ kHz, the programmed all-pass filter resulted in

$$G_{\mathrm{AP}}(z) = \frac{0.8816\, z^2 - 1.8779\, z + 1}{z^2 - 1.8779\, z + 0.8816}. \tag{21}$$

Gains k_{pm} and k_{vm} must fit the power and voltage oscillations to the interval $[-1,1]$, resulting $k_m \equiv k_{pm} \cdot k_{vm}$ in Equation (14). Values $k_{pm} = 0.5$ and $k_{vm} = 2$ are used to set a high δ sensitivity, while ensuring the non-saturation of δ when operating at the neighbourhoods of the MPP.

The condition $i_{\mathrm{pv}} \leq i_{\mathrm{start}}$ in Figure 9 inhibits the MPPT at the start-up, where the duty-cycle is small and operation is in open-circuit with $i_{\mathrm{pv}} = 0$, and hence without any chance to get information by power modulation. Thus, the converter starts on the right side of the MPP with $\delta = 1$, i.e., with a conventional PI action.

On the other hand, when the current error becomes negative, MPPT_ON is set to zero and $\delta = 1$.

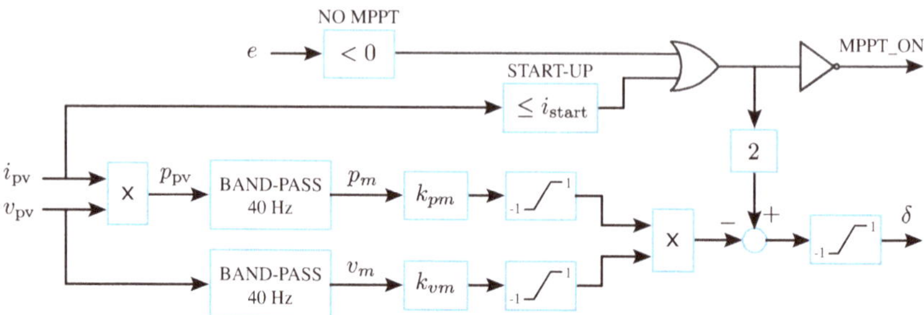

Figure 9. Detection of the operating point position relative to the MPP using the parameter δ.

Figure 10 shows the proposed modified PI current control including MPPT action. As mentioned, if $e < 0$, then $\delta = 1$ is applied, the duty-cycle modulation stops, and thus the current control is a conventional PI control. On the contrary, when $e > 0$, the duty-cycle modulation is initiated and δ is calculated. The current error e is limited to 1 A, so that the speed of convergence to the MPP is given by the integrator gain k_{i_I} and the value of δ, but it is not dependent on the current reference level. As the converter approaches the MPP, δ tends to zero and the PI slows down the duty-cycle variation to finally get a still MPP operation.

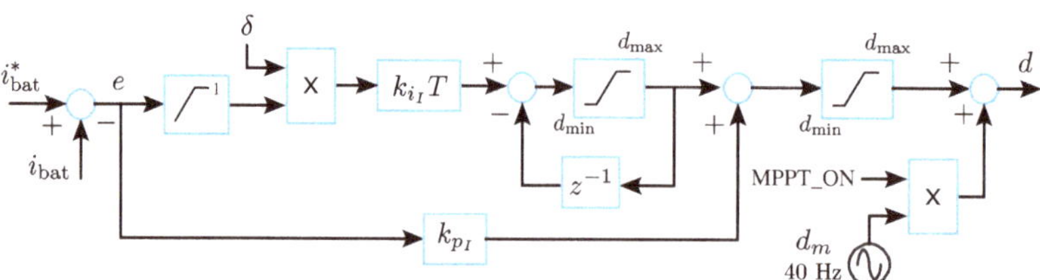

Figure 10. Proposed proportional-integral (PI) current control modification to achieve In-Cond MPPT.

The proposed modified PI has to ensure the stability and fast response of the control shown in Figure 6a for all operating points on the right side of the MPP. The design worst case is with $\delta_{dc} = 1$, that is, when MPPT action is inhibited and the controller behaves as a PI compensator. Expressing the PI in its continuous form

$$PI(s) = k_{p_I} + \frac{k_{i_I}}{s} = k_{p_I} \cdot \frac{s + \omega_z}{s}, \tag{22}$$

the zero ω_z is designed at the minimum value of the natural frequency

$$\omega_z = \omega_{n_{\min}} \approx \frac{V_o}{V_{\text{MPP}} \sqrt{LC}} \tag{23}$$

in order to get the maximum phase margin as possible.

The gain k_{p_I} is designed to achieve a high control bandwidth by setting the loop-gain crossover frequency ω_c at $\omega_{\text{nyq}}/6 = \pi/(6T)$. At these high frequencies, the duty-to-current transfer function (4) can be approximated by

$$G_i(s) \approx \frac{k_i \omega_n^2}{\omega_{z_i} s} = \frac{V_{\text{pv}}}{Ls} \tag{24}$$

that exhibits the maximum gain (worst case) when operating at the open-circuit voltage $V_{\text{pv}} = V_{\text{oc}}$ and with all three converters working in parallel $L = L_j/3$. Hence, a simple design equation for k_{p_I} yields $1 = |PI(j\omega_c)| \cdot V_{\text{oc}}/(L\omega_c)$, or

$$k_{p_I} = \frac{L\omega_c^2}{V_{\text{oc}} \cdot |j\omega_c + \omega_z|} \tag{25}$$

and

$$k_{i_I} = k_{p_I} \cdot \omega_z \tag{26}$$

The designed values for k_{p_I} and k_{i_I} are given in Table 1.

Table 1. Control parameters.

Description	Variable	Value
Control sampling frequency	$f_s = \frac{1}{T}$	4 kHz
Duty-cycle modulation frequency	f_m	40 Hz
Duty-cycle modulation amplitude	$d_{m_{pk}}$	0.5%
Absorption voltage reference	v_{bat}^*	14.7 V/battery
Float voltage reference	v_{bat}^*	13.6 V/battery
Proportional gain - voltage loop	k_{p_V}	0.05
Integral gain - voltage loop	k_{i_V}	50 rad/s
Current limit	$i_{\max}$	60 A
Proportional gain - current loop	k_{p_I}	0.001
Integral gain - current loop	k_{i_I}	1 rad/s
Minimum PV current for MPPT start-up	i_{start}	50 mA
AC-power gain	k_{pm}	0.5
AC-voltage gain	k_{vm}	2
Bandpass filters, center frequency	f_0	40 Hz
Bandpass filters, bandwidth	f_{BW}	80 Hz

Figure 12 shows the resulting Bode diagrams of the open-loop gain at the two ending points of the stable region ($V_{\text{pv}} = V_{\text{MPP}}$ and $V_{\text{pv}} = V_{\text{oc}}$), where the gain crossover frequency ω_c and both phase and gain margins are detailed. In Figure 12a, the converter operates at the open-circuit voltage with a control bandwidth $\omega_c = 973$ rad/s that ensures a fast response. However, when the converter reaches the MPP in Figure 12b, the control bandwidth is strongly reduced to $\omega_c = 0.1$ rad/s, much lower than the modulation frequency ω_m, and therefore the MPP operation is not affected by the modulation and remains constant without oscillations. The gain and phase margins are large enough to ensure a robust stability in the whole operating range $V_{\text{MPP}} \leq V_{\text{pv}} \leq V_{\text{oc}}$

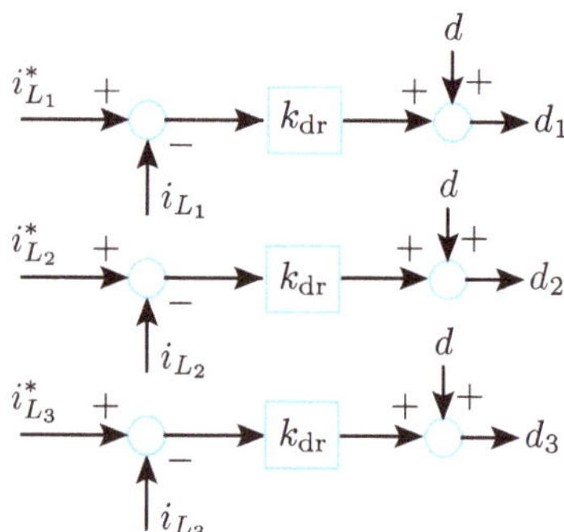

Figure 11. Droop correction to equalize and control output currents of all converters.

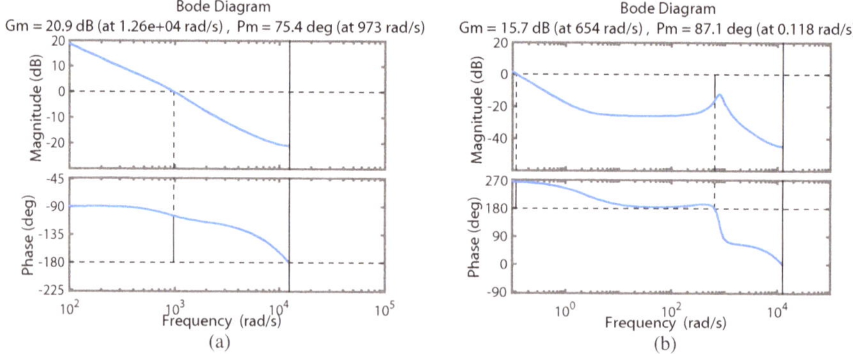

Figure 12. Open-loop Bode diagrams when operating at: (**a**) $V_{\mathrm{pv}} = V_{\mathrm{oc}}$; and (**b**) $V_{\mathrm{pv}} = V_{\mathrm{MPP}}$.

Simulated results are presented in Figure 13, obtained using PSIM$^{©}$, where steps in the charging current reference are applied from 0 A to 20 A. The converter moves from the open-circuit voltage to the MPP in less than 100 ms. The calculated PV efficiency is $\eta = 100 \cdot 422.6/422.9 = 99.9\%$. The ac components p_m and v_m extracted by the band-pass filters and the parameter δ are also shown. The power p_m oscillates at frequency ω_m when operating out of the MPP, but at $2\omega_m$ when operation is at the MPP.

Despite a duty-cycle, d is calculated to maximize the extracted PV power if needed, it is not advisable to directly apply it to all parallelized converters, since they have slight differences in the inductors series resistances r_j and turn-on/off delays, which may cause the currents to unbalance. Instead, a droop strategy is proposed as shown in Figure 11 to distribute duty-cycles d_j to the converters. If the j-th converter is active, its duty-cycle is calculated as

$$d_j = d + \Delta d_j ; \quad \Delta d_j = k_{dr} \cdot (i_{L_j}^* - i_{L_j}), \tag{27}$$

where $i_{L_j}^* = i_{\mathrm{bat}}/n$ is the current reference, n is the number of active converters and $i_{\mathrm{bat}} = \sum i_{L_k}$. On the contrary, when a converter is not active, all transistors Q_j, $Q_{R_{j1}}$ and $Q_{R_{j2}}$ in Figure 1 are switched off and therefore i_{L_j} is zero. It can be noticed that the averaged value of all applied duty-cycles to the active converters is d

$$\sum_j^n \Delta d_j = 0 ; \quad \sum_j^n d_j = nd. \tag{28}$$

Each active converter current satisfies

$$i_{L_j} = (d_j v_{pv} - V_o) G_{ind} \tag{29}$$

and adding for all active converters

$$i_{bat} = \sum_j^n i_{L_j} = (\sum_j^n d_j v_{pv} - n V_o) G_{ind} = (d v_{pv} - V_o) n G_{ind}, \tag{30}$$

which shows, as mentioned before, that the parallel configuration behaves as a single stage with an averaged duty-cycle d and with all inductances G_{ind} in parallel.

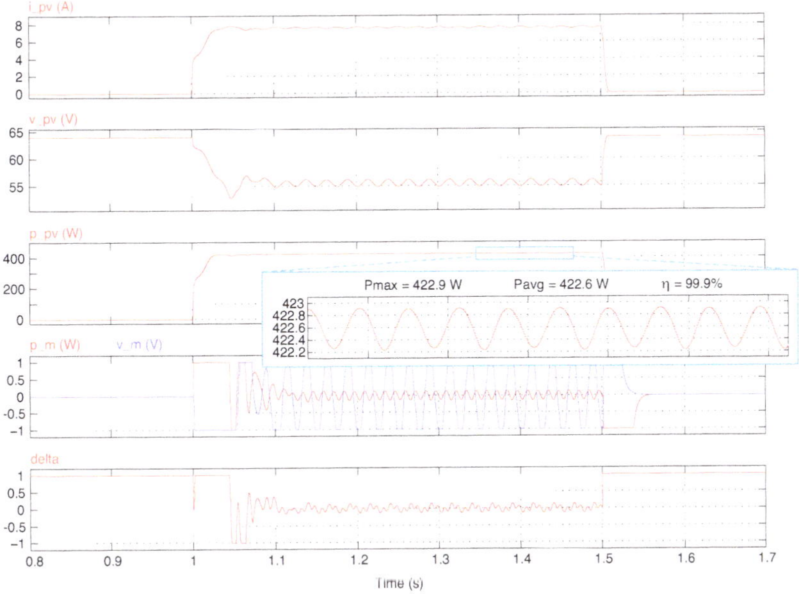

Figure 13. Simulated transient response against current reference steps. The power alternates between zero and the MPP—from top to bottom: i_{pv}, v_{pv}, p_{pv}, p_m, v_m and δ.

Equation (29) in steady-state results $i_{L_j} = (d_j v_{pv} - V_o)/r_j$, and therefore a variation in the duty-cycle Δd_j generates a variation in the converter current $\Delta i_{L_j} = \Delta d_j v_{pv}/r_j$, which gives an estimation for the droop gain k_{dr} in (27). In order to guarantee the currents compensation, the droop gain is designed as

$$k_{dr} \gtrsim \frac{\max r_j}{V_{MPP}}. \tag{31}$$

The proposed droop strategy allows for controlling each converter current individually to optimize the overall efficiency. For instance, when the power is lower than one third of the total installed power, only one converter is active and the other two are kept off, so that switching losses are minimized. When power is between one third and two thirds of the total power, two converters are active and share the power from 50% to 100% of their rated power. Finally, when the power is higher than two thirds of the total, all three converters are activated and share the power from 66% to 100% of their rated power. Moreover, a rotation strategy is also implemented to alternate the active and inactive converters to equalize transistors aging and to minimize thermal cycling. It will be shown in the experimental results that the converters' activation and deactivation for losses rotation does not have a transient effect on the MPP operation, and hence it can be done without affecting the photovoltaic efficiency.

5. Experimental Results

The presented MPPT strategy with droop was assayed in the battery charger shown in Figure 1, whose main parameters are specified in Table 2. Transistors Q_j (j = 1, 2, 3) are fired at f_{sw} = 40 kHz with complementary drive for the rectification transistors $Q_{R_{j1}}$ and $Q_{R_{j2}}$. Schottky diodes D_j drive only during the PWM dead-time. The anti-series transistors $Q_{R_{j2}}$ impede the conduction of the lossy body-diodes of $Q_{R_{j1}}$ during the dead-time. The blocking transistors Q_B are always on, except if the input voltage gets close to the battery voltage or a if a panel reverse current is detected. The converter was designed to charge lead acid batteries with voltage ranging from 12 V to 48 V and charging current up to 60 A. Though the converter is 3 kW rated, presented experimental results were obtained at a lower power, using the 480 W E4350B solar array simulator from Agilent/Keysight Technologies[©].

The control is resolved in a RX630 microcontroller from Renesas Electronics[©] at 4 kHz, using three independent PWM outputs to drive the three converters, and six analog input channels for the PV voltage and current, the battery voltage and the three output currents. A human interface device (HID) class USB communication, readable with computers, tablets, etc., has also been implemented to send internal data at a speed of 2 kB/s.

Table 2. Power converter parameters.

Description	Variable	Value
Nominal power	P	3 kW
Output voltage	V_o	12 V–48 V
Maximum output current	i_{bat}	3×20 A = 60 A
Switching frequency	f_{sw}	40 kHz
Output filter inductances	L_j	130 µH
Inductor series resistances	r_j	25 mΩ
Input capacitance	C	2300 µF

Figure 14 shows the P–V irradiance curves programmed in the E4350B to test the performance of the proposed MPPT method. Initially, the converter is turned on with irradiance level corresponding to curve PV1, which is maintained for five seconds. Then, the irradiance is suddenly changed to the curve PV2 and is maintained for another five seconds, and so on with curves PV3 and PV4. Finally, the converter is turned off with irradiance PV4. Red circles indicate the converter operating point motion and were obtained in real time via USB with a 2 ms sampling period. It can be seen that there is only one red circle in the transition between two consecutive MPPs, which indicates that the MPPT takes less than 4 ms to converge when the irradiance decreases.

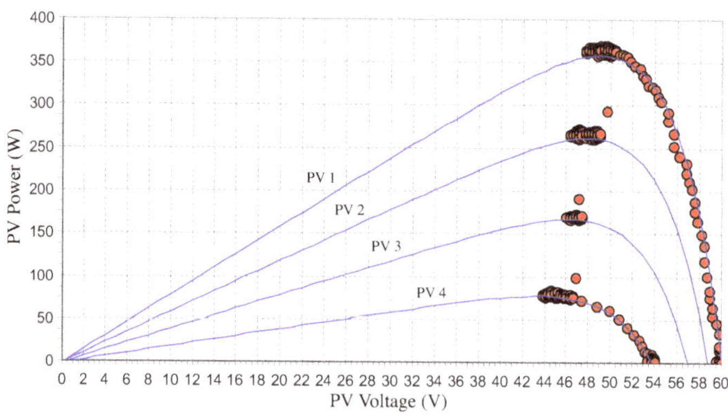

Figure 14. Operating point shift during irradiance step changes from PV1 to PV4 (samp. time = 2 ms).

In more detail, Figures 15 and 16 show transients produced by a sudden increase and decrease in irradiance, respectively. In Figure 15, the irradiance is changed from PV4 to PV1. The new MPP is reached in approximately 50 ms. The 40 Hz oscillation is barely distinguishable in the PV current or voltage, and it cannot be observed in the PV power. In Figure 16, the irradiance is decreased from PV1 to PV4, and, as mentioned before, most of the power transition takes less than 4 ms.

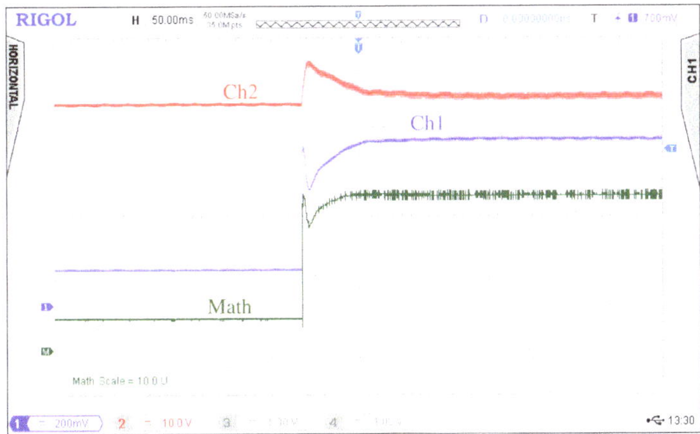

Figure 15. Transient response against an irradiance step from PV4 to PV1. Ch1: PV current (2 A/div). Ch2: PV voltage (10 V/div). Math: PV power (100 W/div).

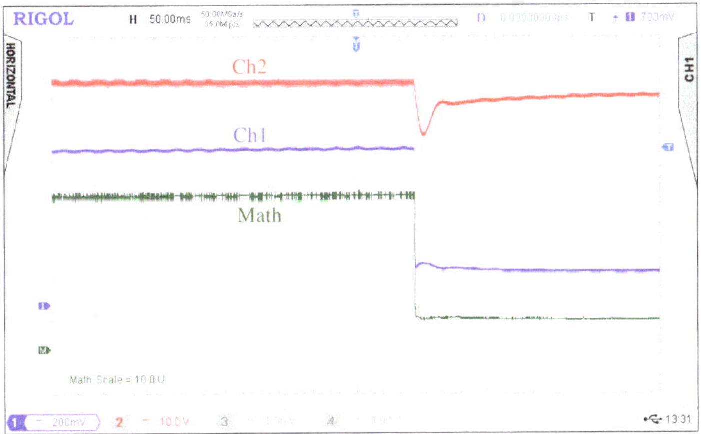

Figure 16. Transient response against an irradiance step from PV1 to PV4. Ch1: PV current (2 A/div). Ch2: PV voltage (10 V/div). Math: PV power (100 W/div).

An irradiance change is not the most challenging case in terms of speed response, since the PV voltage and duty-cycle variations are not wide. However, a large step change in the demanded battery charging power requires a significant variation of the duty-cycle and PV voltage, and this is the worst case for the converter settling time. In this sense, Figure 17 shows the transients produced by steps in i_{bat}^* between 0 A and 20 A when the converter operates at the irradiance PV1 curve. The PV power changes from zero to the MPP, showing a rising time of approximately 100 ms and a falling time around 20 ms. It can be noticed again that the power at the MPP is constant without oscillations.

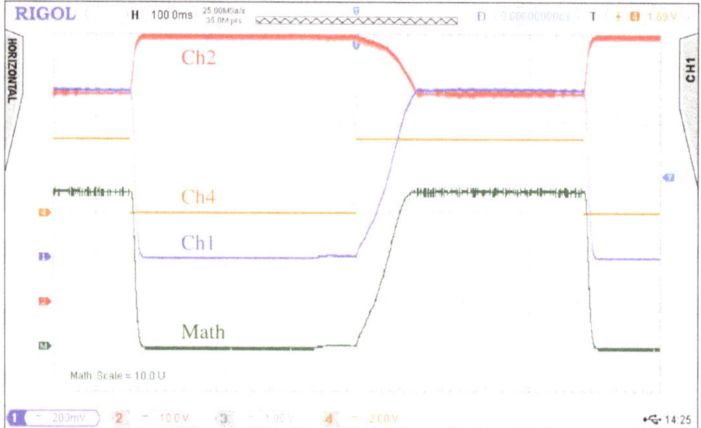

Figure 17. Transient response against a current reference i^*_{bat} steps between 0 A and 20 A. Ch1: PV current (2 A/div). Ch2: PV voltage (10 V/div). CH4: current reference synchronism digital output. Math: PV power (100 W/div).

Finally, Figures 18 and 19 are intended to show the performance of the droop compensation and the on/off switching of parallelized converters. These figures were obtained when charging batteries at 17 A and 28 V giving the maximum 480 W of the solar simulator. At the beginning of Figure 18, the charge current shown in Ch3 is equally shared by converters 1 and 2 by means of droop action. Then, the current reference of converter 1 is set to zero and the reference of converter 2 is set to the total charge current. After around 2 ms, all of the charge current is provided by converter 2, and the converter 1 is turned off. Moreover, Figure 19 shows the effect of a converter activation. Initially, converter 1 is off and all charge currents are provided by converter 2. Then, converter 1 is turned on and current references of both converters are set to half the charge current. Finally, after 2 ms, the current is equally shared by converters 1 and 2. It can be seen that activation and deactivation of converters have no effect on the charging current and hence do not affect the MPP operation.

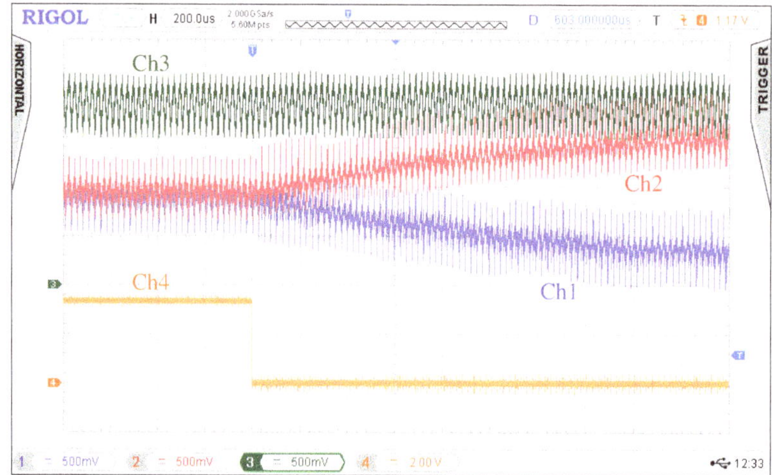

Figure 18. Droop equalization and converter 1 shut-down. Ch1: converter 1 output current (5 A/div). Ch2: converter 2 output current (5 A/div). Ch3: total output current (5 A/div). Ch4: synchronism digital output.

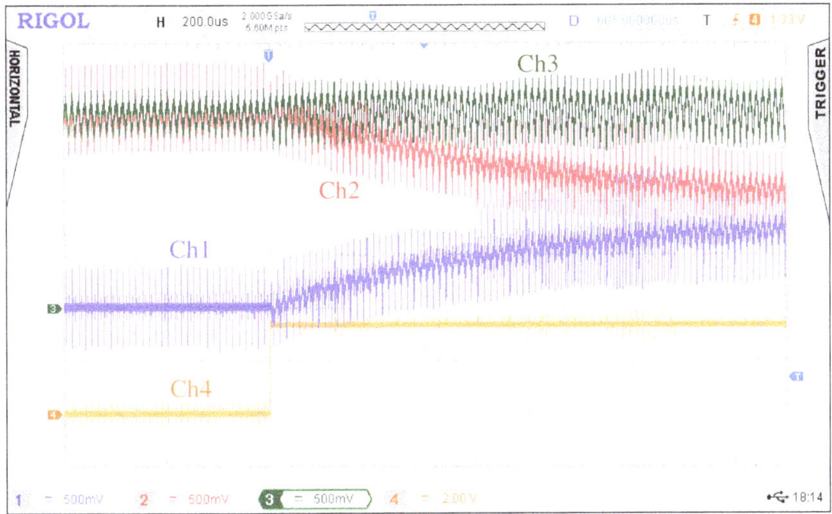

Figure 19. Converter 1 turn-on and current sharing with converter 2 using droop correction. Ch1: converter 1 output current (5 A/div). Ch2: converter 2 output current (5 A/div). Ch3: total output current (5 A/div). Ch4: synchronism digital output.

6. Conclusions

This paper presents a new fast MPPT method for step-down photovoltaic (PV) battery chargers. The method adds a low frequency modulation to the duty-cycle to calculate the conductance error, which is used as a gain in the current loop. Therefore, it can be considered as another MPPT variant of the incremental conductance (IC) type. This produces a quiescent maximum power point (MPP) since the control bandwidth tends to zero at the MPP, yielding PV efficiencies higher than 99%. However, when the operating point is not close to the MPP, the bandwidth of the current control is around 1 krad/s, resulting in a fast convergence to the MPP.

Furthermore, if demanded power is lower than the available maximum PV power, the proposed design ensures a fast regulation on the right side of the MPP. The presented MPPT exhibits, in the worst case, settling times of 50 ms against irradiance changes, and 100 ms against power reference changes.

In addition, the control problem of a parallel arrangement of converters is solved by means of a droop law. The MPPT algorithm gives an averaged duty-cycle for all active converters, and the droop compensation allows duty-cycles to be distributed to all active converters to control their currents individually. Moreover, the droop strategy allows activation and deactivation of converters without affecting the MPP and battery charging operation.

Finally, it is worth noticing that the proposed battery charger control can be solved at low sampling rates using a low-cost microcontroller.

Author Contributions: J.M.E. proposed the main idea, performed the theoretical analysis and wrote the paper. All authors contributed to the practical implementation and experimental validation, paper review and editing. All authors have read and approved the final manuscript.

Funding: This research was funded by the Centro para el Desarrollo Technological Industrial (CDTI) under the project grant number IDI-20160123.

Acknowledgments: We thank the company Dismuntel S.A.L. for its help in building the PV charger prototype.

Conflicts of Interest: The authors declare no conflict of interest.

References

1. Ahmed, J.; Salam, Z. A Modified P&O Maximum Power Point Tracking Method With Reduced Steady-State Oscillation and Improved Tracking Efficiency. *IEEE Trans. Sustain. Energy* **2016**, *4*, 1506–1515.
2. Sher, H.A.; Rizvi, A.A.; Addoweesh, K.E.; Al-Haddad, K. A Single-Stage Stand-Alone Photovoltaic Energy System With High Tracking Efficiency. *IEEE Trans. Sustain. Energy* **2017**, *8*, 755–762. [CrossRef]
3. Metry, M.; Shadmand, M.B.; Balog, R.S.; Abu-Rub, H. MPPT of Photovoltaic Systems Using Sensorless Current-Based Model Predictive Control. *IEEE Trans. Ind. Appl.* **2017**, *53*, 1157–1167. [CrossRef]
4. Escobar, G.; Pettersson, S.; Ho, C.N.M.; Rico-Camacho, R. Multisampling Maximum Power Point Tracker (MS-MPPT) to Compensate Irradiance and Temperature Changes. *IEEE Trans. Sustain. Energy* **2017**, *8*, 1096–1105. [CrossRef]
5. Macaulay, J.; Zhou, Z. A Fuzzy Logical-Based Variable Step Size P&O MPPT Algorithm for Photovoltaic System. *Energies* **2018**, *11*, 1340.
6. Galiano, I.; Ordonez, M. PV Energy Harvesting Under Extremely Fast Changing Irradiance: State-Plane Direct MPPT. *IEEE Trans. Ind. Electron.* **2019**, *66*, 1852–1861.
7. Paz, F.; Ordonez, M. High-Performance Solar MPPT Using Switching Ripple Identification Based on a Lock-In Amplifier. *IEEE Trans. Ind. Electron.* **2016**, *63*, 3595–3604. [CrossRef]
8. Li, C.; Chen, Y.; Zhou, D.; Liu, J.; Zeng, J. A High-Performance Adaptive Incremental Conductance MPPT Algorithm for Photovoltaic Systems. *Energies* **2016**, *9*, 288. [CrossRef]
9. Li, X.; Wen, H.; Jiang, L.; Xiao, W.; Du, Y.; Zhao, C. An Improved MPPT Method for PV System With Fast-Converging Speed and Zero Oscillation. *IEEE Trans. Ind. Appl.* **2016**, *52*, 5051–5064. [CrossRef]
10. Nguyen-Van, T.; Abe, R.; Tanaka, K. MPPT and SPPT Control for PV-Connected Inverters Using Digital Adaptive Hysteresis Current Control. *Energies* **2018**, *11*, 2075. [CrossRef]
11. Brundlinger, R.; Henze, N.; Haberlin, H.; Burger, B.; Bergmann, A.; Baumgartner, F. prEN 50530—The New European standard for performance characterisation of PV inverters. *Eur. Photovolt. Sol. Energy Conf.* **2009**, 3105–3109.

Article

A Method to Enhance the Global Efficiency of High-Power Photovoltaic Inverters Connected in Parallel

Marian Liberos *, Raúl González-Medina, Gabriel Garcerá and Emilio Figueres

Grupo de Sistemas Electrónicos Industriales del Departamento de Ingeniería Electrónica,
Universitat Politècnica de València, Camino de Vera s/n, 46022 Valencia, Spain; raugonme@upv.es (R.G.-M.);
ggarcera@eln.upv.es (G.G.); efiguere@eln.upv.es (E.F.)
* Correspondence: malimas@upv.es; Tel.: +34-963-879-606

Received: 15 May 2019; Accepted: 6 June 2019; Published: 11 June 2019

Abstract: Central inverters are usually employed in large photovoltaic farms because they offer a good compromise between costs and efficiency. However, inverters based on a single power stage have poor efficiency in the low power range, when the irradiation conditions are low. For that reason, an extended solution has been the parallel connection of several inverter modules that manage a fraction of the full power. Besides other benefits, this power architecture can improve the efficiency of the whole system by connecting or disconnecting the modules depending on the amount of managed power. In this work, a control technique is proposed that maximizes the global efficiency of this kind of systems. The developed algorithm uses a functional model of the inverters' efficiency to decide the number of modules on stream. This model takes into account both the power that is instantaneously processed and the maximum power point tracking (MPPT) voltage that is applied to the photovoltaic field. A comparative study of several models of efficiency for photovoltaic inverters is carried out, showing that bidimensional models are the best choice for this kind of systems. The proposed algorithm has been evaluated by considering the real characteristics of commercial inverters, showing that a significant improvement of the global efficiency is obtained at the low power range in the case of sunny days. Moreover, the proposed technique dramatically improves the global efficiency in cloudy days.

Keywords: efficiency improvement; photovoltaic inverters; parallel inverters

1. Introduction

Photovoltaic (PV) generation has had rapid growth in the last years and is now a significant contribution to the renewable sources of electricity [1–5]. With the purpose of improving the profitability of photovoltaic systems, large-scale PV plants are being installed [6]. In large PV plants, low-power decentralized architectures based on string inverters are usually avoided due to their high cost. Therefore, photovoltaic farms are usually connected to the grid through central inverters that manage the whole power of the system, since they offer a good compromise between costs and efficiency [7].

Figure 1 shows two alternatives to build up a centralized inverter that connects a large PV field to the distribution grid. The efficiency of central inverters composed by a single power stage, Figure 1a, is poor in the low power range when the radiation conditions are low. In the range of MWs, the scheme showed by Figure 1b is preferred, as the parallel connection of several modules offers redundancy, scalability, and a certain degree of fault tolerance. A popular technique to manage the connection of paralleled inverters is average current-sharing (CS). CS offers several advantages, such as a good power sharing among the modules and simplicity of implementation. However, the efficiency of the whole system with this technique at low power is not improved with regard to the single-module configuration [6,8].

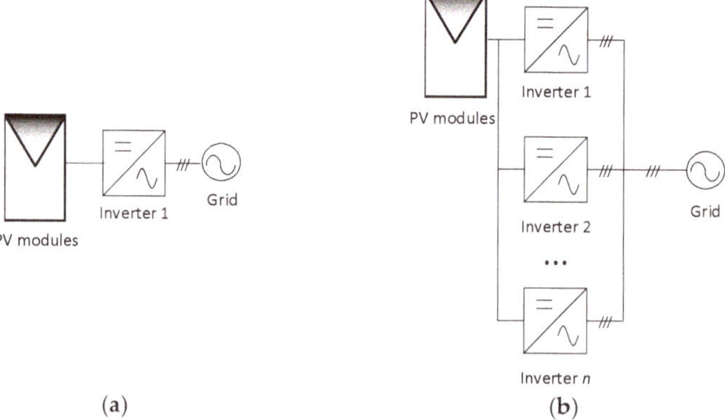

Figure 1. Topologies of high power central inverters. (**a**) Centralized inverter composed by a single power stage; (**b**) centralized inverter composed by n parallel modules.

To overcome this problem, the inverter modules can be connected and disconnected depending on the global delivered power. The concept of connecting/disconnecting the phases of a converter depending on the load-current level has been also applied in several works to low power converters [9–13]. In [9] the phase shedding points are calculated based on a lookup table defined by the junction temperature and on-resistance of MOSFETs. In [10] a multiphase buck converter with a rotating phase-shedding scheme has been presented. In [11] a method is proposed that linearly increases/reduces the power delivered by some channels when the demanded power changes. In [13] a time optimal digital controller for the phase shedding in multiphase buck converters has been developed. Although all these methods are based on the connection/disconnection of phases (or modules) to improve the efficiency of the global system, they cannot be directly extrapolated to high-power inverters.

Some studies regarding the connection/disconnection of paralleled inverters have been developed in the past. In [14] a methodology based on unidimensional efficiency curves and a genetic algorithm was presented. In that work, a unidimensional model is used, so the changes in the maximum power point tracking (MPPT) voltage are not considered. Therefore, this algorithm could be inappropriate in most of the photovoltaic applications. Moreover, the stochastic nature of the genetic algorithm requires the implantation of complex processes to obtain a useful result. This fact could impede the real-time implementation of the technique. In [15], a piecewise curve fitting is used to define the efficiency function and an artificial-intelligence-based algorithm is implemented to obtain the optimized current-sharing. As in [15], a unidimensional model is used to calculate the efficiency of the inverter, and the algorithm proposes a random initialization that requires many iterations to obtain a valid result. Finally, in [16], each converter regulates its respective output power following an algorithm of prioritization. However, the algorithm improves the efficiency in the low range of power but worsens the efficiency in the middle and high power range.

Regarding the efficiency of inverters, there are several efficiency models for photovoltaic inverters that have been proposed in the literature. These models can be classified as unidimensional and bidimensional, depending on whether they only take into account the generated power, or if they consider both the power generation and the DC voltage at the input of the inverter. Some unidimensional and bidimensional models were studied in [17–25].

In this work, a control technique is proposed that decides, in real time conditions, the proper number of inverters that should be on stream to improve the global efficiency in the whole power range. The developed algorithm is based on a bidimensional model of the inverters' efficiency, which takes into account not only the amount of delivered power but also the value of the DC voltage at the input of the inverters, which is continuously changing to achieve the maximum power point (MPP) of the PV field. The algorithm calculates the efficiency of the whole power system by taking into account various scenarios and selects the one that offers the best instantaneous efficiency. A comparative study of the model's accuracy has also been carried out by considering data of several commercial inverters. Data have been obtained from the "Grid Support Inverters List" published by the California Energy Commission [26]. However, it is worth noting that the parameters could be easily extracted from the datasheet of manufacturers, or obtained by means of a reduced number of efficiency measurements on the inverter.

The main contributions of this paper are the following:

1. A comparative study of the advantages and limitations of various models of efficiency for photovoltaic inverters.
2. The proposal of an algorithm that decides the number of inverter modules that should be on stream to obtain the maximum efficiency of the global system in the whole operation range. An advantage of the proposed algorithm is the low requirements in terms of computational resources so that it can be easily implemented in a real time operation.
3. A detailed study of the expected performance of the proposed algorithm when it is applied to several commercial photovoltaic inverters.

2. Efficiency Modeling of PV Inverters

As it has been pointed out in the previous section, two kinds of functional models have been proposed in the literature to evaluate the efficiency of power inverters: unidimensional, in which only the generated power is considered, and bidimensional, in which takes into account both the generated power and the DC voltage at the input of the inverter, which agrees with the MPPT voltage in the case of central inverters.

2.1. Unidimensional Models

The unidimensional model of Jantsch [17–19] is expressed by (1), where k_0, k_1, and k_2 are the coefficients to be calculated for any inverter, and c is the load factor, which is defined as the ratio between the power that is processed at a certain instant and the nominal power of the inverter. In (1), the part of the losses that are independent of the generated power (constant losses) are weighted by k_0, the losses that linearly depend on the load factor are weighted by k_1 and the losses with a quadratic dependence on the load factor are weighted by k_2.

$$\eta(c) = \frac{c}{c + (k_0 + k_1 c + k_2 c^2)} \tag{1}$$

Dupont [20] indicates that the efficiency of power inverters can be approximated by the second order function (2), being α_1, α_0, β_1, and β_0 coefficients that can be obtained by applying curve fitting algorithms over experimental measurements and c is the load factor of the inverter.

$$\eta(c) = \frac{\alpha_1 c + \alpha_0}{c^2 + \beta_1 c + \beta_0} \tag{2}$$

2.2. Bidimensional Models

In photovoltaic inverters that are directly connected to the PV field, as is the case of central inverters, the DC input voltage of the inverters is continuously following the operation point that is calculated by the maximum power point tracking (MPPT) algorithm. Therefore, the use of unidimensional models that consider the DC voltage as constant is inappropriate to properly evaluate the actual efficiency of the inverter in the whole operation range. To overcome this problem, several bidimensional models have been proposed.

Rampinelli [23] modifies the model represented by (1), which only considers the efficiency as a function of the delivered power, by taking also into account the influence of the input voltage on the predicted efficiency. To achieve this, the coefficients k_0, k_1, and k_2 are expressed as a function of the input voltage being modified as $k_0'{}_{(vin)}$, $k_1'{}_{(vin)}$, and $k_2'{}_{(vin)}$. The expressions of these coefficients are defined as (3)–(5), by assuming that the coefficients have a linear dependency with the input voltage and $k_{0,0}$, $k_{0,1}$, $k_{1,0}$, $k_{1,1}$, $k_{2,0}$, and $k_{2,1}$ are the coefficients to be calculated. This modeling approach is described by (6).

$$k_0'(v_{in}) = K_{0,0} + k_{0,1}v_{in} \tag{3}$$

$$k_1'(v_{in}) = K_{1,0} + k_{1,1}v_{in} \tag{4}$$

$$k_2'(v_{in}) = K_{2,0} + k_{2,1}v_{in} \tag{5}$$

$$\eta(c,\, v_{in}) = \frac{c}{c + \left(k_0'(v_{in}) + k_1'(v_{in})c + k_2'(v_{in})c^2\right)} \tag{6}$$

Similarly, if it is assumed that the efficiency can vary in a quadratic way regarding both the delivered power and the DC voltage, the coefficients can be calculated as (7)–(9). As a result, Equation (10) describes a model of the inverter that considers a nonlinear dependency with the input voltage.

$$k_0''(v_{in}) = K_{0,0} + k_{0,1}v_{in} + k_{0,2}v_{in}^2 \tag{7}$$

$$k_1''(v_{in}) \doteq K_{1,0} + k_{1,1}v_{in} + k_{1,2}v_{in}^2 \tag{8}$$

$$k_2''(v_{in}) = K_{2,0} + k_{2,1}v_{in} + k_{2,2}v_{in}^2 \tag{9}$$

$$\eta(c,\, v_{in}) = \frac{c}{c + \left(k_0''(v_{in}) + k_1''(v_{in})c + k_2''(v_{in})c^2\right)} \tag{10}$$

In [24] Sandia Laboratories propose a mathematical model that describes the performance of inverters. The Sandia model is represented in (11)–(14), where pac is the output power of the inverter; p_{ac_o} is the nominal AC power rating, pdc is the input power of the inverter, vdc is the input voltage of the inverter, vdc_o is the nominal DC voltage, pdc_o is the nominal input power, pso is the minimum considered DC power; co is a parameter that defines the curvature of the relationship between AC power and DC power at nominal voltage. Finally, $c1$, $c2$, and $c3$ are coefficients that represent the linear relationship between p_{dc_o}, p_{so}, and c_o, respectively, and the DC input voltage.

$$A = p_{dc\_o}\left(1 + c_1(v_{dc} - v_{dc\_o})\right) \tag{11}$$

$$B = p_{so}\left(1 + c_2(v_{dc} - v_{dc\_o})\right) \tag{12}$$

$$C = c_o\left(1 + c_3(v_{dc} - v_{dc\_o})\right) \tag{13}$$

$$p_{ac} = \left(\frac{p_{ac\_o}}{A - B} - C(A - B)\right)(p_{dc} - B) + C(p_{dc} - B)^2 \tag{14}$$

$$\eta = \frac{p_{ac}}{p_{dc}} \tag{15}$$

Finally, Driesse [25] proposes a model defined as (16)–(19), with $b_{0,0\ldots2}$, $b_{10\ldots2}$, $b_{2,0\ldots2}$ being the coefficients to be fitted.

$$b_0(v_{in}) = b_{0,0} + b_{0,1}(v_{in} - 1) + b_{0,2}\left(\frac{1}{v_{in}} - 1\right) \tag{16}$$

$$b_1(v_{in}) = b_{1,0} + b_{1,1}(v_{in} - 1) + b_{1,2}\left(\frac{1}{v_{in}} - 1\right) \tag{17}$$

$$b_2(v_{in}) = b_{2,0} + b_{2,1}(v_{in} - 1) + b_{2,2}\left(\frac{1}{v_{in}} - 1\right)$$ (18)

$$\eta(c, v_{in}) = \frac{c}{c + b_0(v_{in}) + b_1(v_{in})c + b_2(v_{in})c^2}$$ (19)

3. Description of the Proposed Efficiency Oriented Algorithm

As it was described in the introduction section, the proposed algorithm calculates the optimal number of parallel modules of a central inverter that should be on stream to maximize the global efficiency in the whole power range.

The algorithm is initially based on the calculation of the local maxima by applying the second derivative test to the function that predicts the efficiency of the whole system, $eff(c_i, v_{in})$. This function (20) computes the efficiency of the whole system starting from the efficiency of each one of the modules, $\eta(c_i, v_{in})$ that can be obtained by means of one of the functional models that were described in the previous section. In (20), c_i $(i = 1, 2, \ldots, n)$ is the load factor of each parallel inverter, i.e., the ratio between the power that is actually managed by each module and its nominal power. The DC voltage of the central inverter is represented by v_{in}.

$$eff(c_i, v_{in}) = \sum_{i=1}^{n} \frac{c_i}{c_1 + c_2 + \ldots + c_n} \cdot \eta(c_i, v_{in})$$ (20)

To apply the second derivative test, the critical points of the function (20) can be calculated by solving the equations' system obtained from the first partial derivatives. From the second derivative, the Hessian matrix (21) can be obtained [27]. Finally, if H in a critical point is negative definite, that critical point is a local maximum. Therefore, the optimal load factor for each one of the modules on stream that maximizes the global efficiency can be obtained by solving (20) and (21) for a certain operating point described by both the DC voltage and the supplied power. It is worth pointing out that the second derivative test does not calculate the maxima points when some of $c_i = 0$. To solve this issue, n different $eff_j(c_i, v_{in})$ can be defined, being $j = 1 \ldots n$ and $i = 1 \ldots n$. Following this procedure, j relative maximums are obtained, one for each $eff_j(c_i, v_{in})$, with the searched maximum being the highest of these.

$$H = \begin{bmatrix} \frac{\partial^2 eff}{\partial c_1^2} & \cdots & \frac{\partial^2 eff}{\partial c_n c_1} \\ \vdots & \ddots & \vdots \\ \frac{\partial^2 eff}{\partial c_1 c_n} & \cdots & \frac{\partial^2 eff}{\partial c_n c_n} \end{bmatrix}$$ (21)

One disadvantage of this procedure is the very high computation time needed to calculate all the critical points for a large set of c_i values. However, all the local maximums for a certain $eff_j(c_i, v_{in})$ are produced when the power is equally shared among the modules, i.e., $c_1 = c_2 \ldots = c_n$. Therefore, the method can be simplified since only the number of modules on stream that maximizes the global efficiency should be calculated. By applying this condition, a practical implementation of the proposed method can be obtained, which is shown in the following section.

Practical Implementation of the Proposed Efficiency Oriented Algorithm

Starting from the predictions of a functional model of inverters efficiency, the algorithm calculates the efficiency of the central inverter by considering all possible combinations of modules on stream and chooses the result that offers the maximum efficiency. As it has been highlighted before, the relative maximum values of efficiency are achieved when the power is equally shared among the modules, so it should be calculated only the value of n that maximizes the global efficiency. One of the most important characteristics of the simplified algorithm is the low need for computational resources and its easy implementation to work in real time conditions.

Figure 2 shows the flowchart of the proposed algorithm. In the figure, v_{in} and p_{in} are the MPPT input voltage and the generated power, respectively, while vdc_{min} and vdc_{max} are the limits of the MPP voltage range of the inverter. The nominal power of the photovoltaic farm and the one of each parallel inverter are represented by p_{tot} and p_{mod}, respectively; n is the total number of parallel modules; n_i, c_{mod-i}, c_{tot-i}, η_{mod-I}, and η_{tot-i} (for $i = 1$ to n) represent the number of inverters considered in each iteration, the load factor of only a module, the load factor of the whole system, the efficiency of each module and the one of the whole system, respectively, in all cases for the corresponding iteration. Finally, n_{ON} is the number of modules on stream that achieves the global maximum efficiency η_{max}.

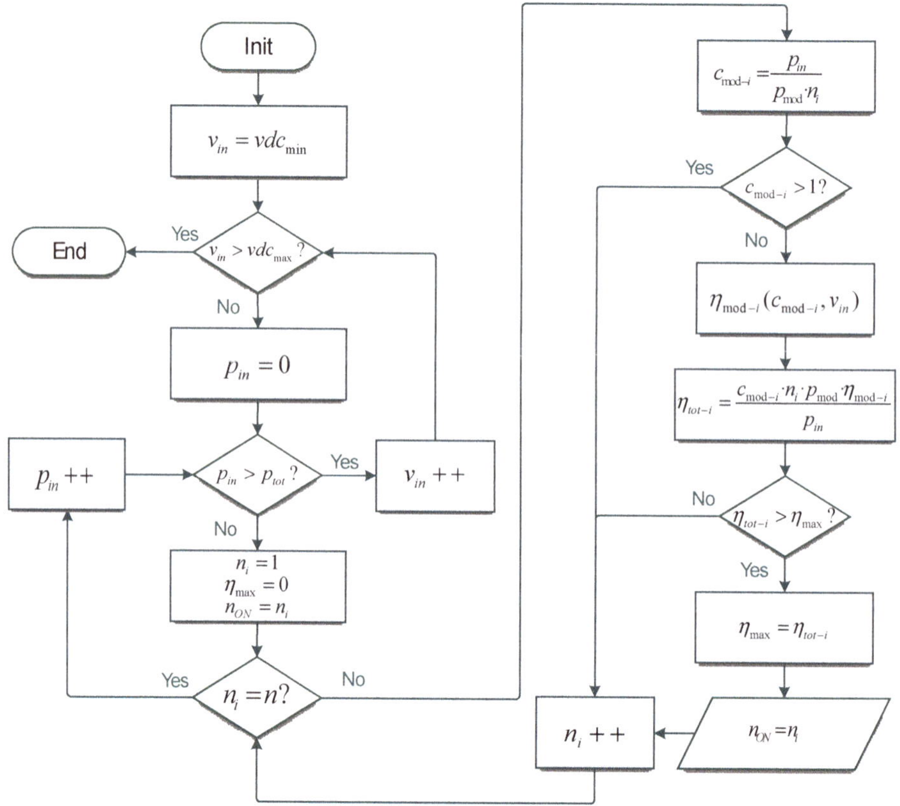

Figure 2. Efficiency-oriented algorithm.

4. Methods

4.1. Selection of Inverters for the Study

Table 1 summarizes the commercial inverters that have been evaluated to validate the proposed concepts. The required data to build up the efficiency models of the inverters have been extracted from the "Grid Support Inverters List" that California Energy Commission (CEC) publishes [26].

Table 1. List of inverters under study (Source: CEC (California Energy Commission) Grid Support Inverters List).

Number of Inverter	Inverter	Company	Nominal Power
1	MPS-250HV	Dynapower	250 kW
2	EQX0250UV480TN	Perfect Galaxy International Ltd.	250 kW
3	FS0501CU	Power Electronics	500 kW
4	IF500TL-UL OUTDOOR	Jema Energy	500 kW
5	XP500U-TL	KACO new energy	500 kW
6	EQMX0500UV320XP	Perfect Galaxy International Ltd.	500 kW
7	SGI 500XTM	Yaskawa Solectria Solar	500 kW
8	PV-625-XRLS-VV-STG	Perfect Galaxy International Ltd.	625 kW
9	Conext Core XC680-NA	Schneider Electric	680 kW
10	EQX0750UV320XP	Perfect Galaxy International Ltd.	750 kW
11	ULTRA-750-TL-OUTD-4-US	Power-One	750 kW
12	SC750CP-US	SMA America	770 kW
13	ULTRA-750-TL-OUTD-1-US	ABB	780 kW
14	HS-P1000GLO-U	Hyosung Heavy Industries	1 MW
15	ULTRA-1100-TL-OUTD-2-US	ABB	1 MW
16	ULTRA-1100-TL-OUTD-4-US	ABB	1 MW
17	EQX1000UV400XP	Perfect Galaxy International Ltd.	1 MW
18	ULTRA-1100-TL-OUTD-4-US	Power-One	1 MW
19	SG1000MX	Sungrow Power Supply	1 MW
20	FS0900CU	Power Electronics	1 MW

Although the proposed methods have been applied to all the inverters summarized by Table 1, in the following only a selection of the most representative results is shown. To choose those representative results, 3 inverters with significant differences in their respective dependence of the MPP voltage on the inverter efficiency have been considered. In Figure 3, the relationship between the input voltage and the efficiency of the listed inverters has been represented. Note that, in some cases, the curves have an ascendant and nonlinear relationship with the input voltage; in some others, they have a descendant and linear relationship with the voltage and, finally, some curves have a descendant and nonlinear relationship with the voltage.

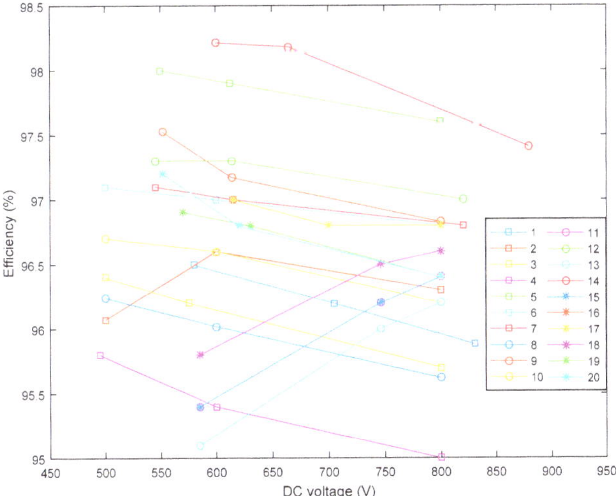

Figure 3. Relationship between the input voltage and the efficiency of the inverters extracted from the CEC Grid Support Inverters List.

To consider the three possibilities (categories) of the efficiency dependency regarding the voltage, a sample of each category for the study presented in this paper has been chosen. Thus, the three chosen inverters have been the following: EQX0250UV480TN (Perfect Galaxy International Ltd.), ULTRA-750-TL-OUTD-4-US (Power-One), and FS0900CU (Power Electronics). Tables 2–4 express the data of the selected inverters that have been extracted from the CEC "Grid Support Inverters List". The intermediate value of v_{in} will be denoted as nominal in the following.

Table 2. Perfect Galaxy International Ltd. EQX0250UV480TN efficiency data.

v_{in}	10%	20%	30%	50%	75%	100%
500 V	94.8%	96.4%	96.8%	97%	96.9%	96.07%
600 V	94.4%	96.1%	96.6%	96.8%	96.8%	96.6%
800 V	93.4%	95.5%	96.1%	96.6%	96.5%	96.3%

Table 3. Power-One ULTRA-750-TL-OUTD-4-US efficiency data.

v_{in}	10%	20%	30%	50%	75%	100%
585 V	94.4%	96%	96.4%	96.5%	96.1%	95.4%
746 V	94.9%	96.5%	96.8%	97.1%	96.8%	96.2%
800 V	95%	96.6%	96.9%	97.1%	96.9%	96.4%

Table 4. Power Electronics FS0900CU efficiency data.

v_{in}	10%	20%	30%	50%	75%	100%
552 V	95.8%	97.5%	97.9%	97.9%	97.7%	97.2%
620 V	95.5%	97.1%	97.5%	97.6%	97.3%	96.8%
800 V	94.8%	96.6%	97.1%	97.2%	96.9%	96.4%

4.2. Modeling of Inverters

In this section, the parameters of the efficiency models presented in Section 2 have been calculated. Tables 5–10 show the coefficients of the models for each inverter under study.

Table 5. Jantsch coefficients.

Coefficient	Perfect Galaxy International Ltd. EQX0250UV480TN	Power-One ULTRA-750-TL-OUTD-4-US	Power Electronics FS0900CU
k_0	0.0044	0.0041	0.0042
k_1	0.016	0.0123	0.0044
k_2	0.0171	0.02245	0.0243

Table 6. Dupont coefficients.

Coefficient	Perfect Galaxy International Ltd. EQX0250UV480TN	Power-One ULTRA-750-TL-OUTD-4-US	Power Electronics FS0900CU
α_0	1.2204	1.2469	0.2538
α_1	46.9698	33.3255	39.5753
β_0	1.5224	1.4687	0.4354
β_1	47.4906	33.5318	39.704

Table 7. Rampinelli coefficients.

Coefficient		Perfect Galaxy International Ltd. EQX0250UV480TN	Power-One ULTRA-750-TL-OUTD-4-US	Power Electronics FS0900CU
$k_0{'}$	$k_{0,0}$	0.0024	0.0055	0.0028
	$k_{0,1}$	3.2176×10^{-6}	-1.9691×10^{-6}	2.1341×10^{-6}
$k_1{'}$	$k_{1,0}$	0.0013	0.0189	-0.0105
	$k_{1,1}$	2.3093×10^{-5}	-9.3061×10^{-6}	2.2688×10^{-5}
$k_2{'}$	$k_{2,0}$	0.0342	0.0526	0.0195
	$k_{2,1}$	-2.6958×10^{-5}	-3.9468×10^{-5}	7.3485×10^{-6}

Table 8. Rampinelli nonlinear model coefficients.

Coefficient		Perfect Galaxy International Ltd. EQX0250UV480TN	Power-One ULTRA-750-TL-OUTD-4-US	Power Electronics FS0900CU
$k_0{''}$	$k_{0,0}$	0.013	0.002	0.0099
	$k_{0,1}$	-3.0294×10^{-5}	8.2289×10^{-6}	-1.9309×10^{-5}
	$k_{0,2}$	2.5508×10^{-8}	-7.4294×10^{-9}	1.5703×10^{-8}
$k_1{''}$	$k_{1,0}$	-0.0961	0.0764	-0.0982
	$k_{1,1}$	3.3111×10^{-4}	-1.8013×10^{-4}	2.8608×10^{-4}
	$k_{1,2}$	-2.3442×10^{-7}	1.2444×10^{-7}	-1.9289×10^{-7}
$k_2{''}$	$k_{2,0}$	0.1967	-0.0461	-0.0343
	$k_{2,1}$	-5.4074×10^{-4}	-3.9468×10^{-5}	-3.7155×10^{-5}
	$k_{2,2}$	3.9098×10^{-7}	-1.4074×10^{-8}	3.259×10^{-8}

Table 9. Sandia coefficients.

Coefficient	Perfect Galaxy International Ltd. EQX0250UV480TN	Power-One ULTRA-750-TL-OUTD-4-US	Power Electronics FS0900CU
p_{ac_o}	250,000	750,000	1.02×10^6
v_{dc_o}	600	746	620
p_{dc_o}	259,520	779,730	1.0524×10^6
p_{so}	1216.1	3395.1	4260.9
c_o	7.8878×10^{-8}	-3.2207×10^{-8}	-2.2411×10^{-8}
c_1	-2.9565×10^{-6}	-4.9517×10^{-5}	3.1507×10^{-5}
c_2	1.1491×10^{-4}	-4.858×10^{-4}	5.7319×10^{-4}
c_3	-0.002	-0.0014	2.9497×10^{-4}

Table 10. Driesse coefficients.

Coefficient		Perfect Galaxy International Ltd. EQX0250UV480TN	Power-One ULTRA-750-TL-OUTD-4-US	Power Electronics FS0900CU
b_0	$b_{0,0}$	-0.302	2.96	0.0755
	$b_{0,1}$	2.4712×10^{-6}	4.465×10^{-6}	2.2934×10^{-6}
	$b_{0,2}$	-0.3054	2.9634	0.073
b_1	$b_{1,0}$	3.9675	7.7702	-22.9922
	$b_{1,1}$	3.283×10^{-5}	7.5838×10^{-6}	-2.8361×10^{-5}
	$b_{1,2}$	3.9788	7.7744	-23.0511
b_2	$b_{2,0}$	17.3126	4.4107	-15.3265
	$b_{2,1}$	1.5323×10^{-5}	-2.9989×10^{-5}	-2.6754×10^{-5}
	$b_{2,2}$	17.3335	4.3712	-15.3923

The coefficients that Tables 5–8 show have been calculated by applying fitting algorithms to the data obtained from the CEC for each inverter under study (Tables 2–4). The fitting algorithms have been applied to Equations (1), (2), (6), and (10) for the Jantsch, Dupont, Rampinelli, and Rampinelli nonlinear models, respectively. To apply the fitting algorithms the Statistics and Machine Learning Toolbox of MATLAB^{TM} has been employed [28].

In Table 9, the coefficients of the Sandia model have been expressed. To obtain the parameters of the Sandia Model (11)–(14), three separate parabolic fits (2nd order polynomial) have been carried out providing the parameters pdc_o, pso, and co for each value of DC voltage. The resultant quadratic formula for each voltage value has been used to obtain p_{so} by solving the x-intercept when $p_{ac} = 0$. In a similar way, p_{dc_o} can be obtained by calculating the x-intercept when $p_{ac} = p_{ac\_o}$. In the model, p_{ac_o} is assumed to be equal to the nominal power of each module and the parameter co has been considered as the second order coefficient obtained in the polynomial fit. The coefficients $c1$, $c2$, and $c3$ have been determined using the p_{dc_o}, p_{so}, and c_o values obtained from the separate parabolic fits. These values are linearly fitted considering their DC voltage dependence. From the resultant equations the coefficients p_{dc_o}, c_1; p_{so}, c_2, and c_o, c_3 have been obtained.

The coefficients of Table 10 have been calculated by applying the fitting algorithms to the data obtained from the CEC and considering Equations (16)–(19).

5. Results

5.1. Evaluation of the Model's Performance

A comparative study of the accuracy of the efficiency models is carried out in this section. The results are compared to the actual CEC measurements to evaluate the proper prediction capability of each model.

Figures 4 and 5 show the efficiency curves that are computed by the unidimensional models Jantsch and Dupont when they are applied to the inverters under study. The CEC data around the fitted curve have been highlighted. As expected, with both models the predicted values cannot be accurate in all the range of the DC voltage.

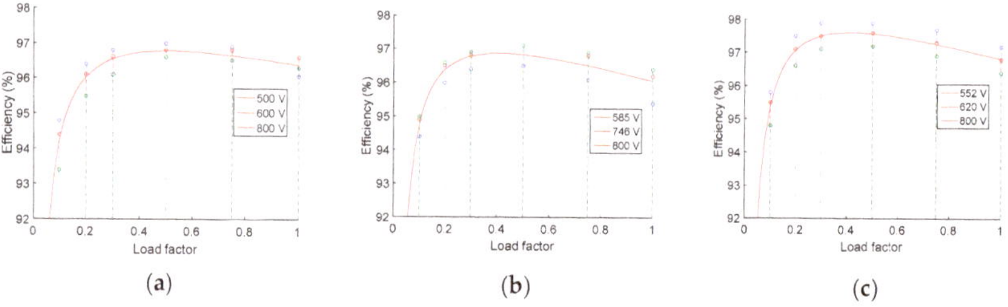

(a)   (b)   (c)

Figure 4. Efficiency curves calculated by means of the Jantsch model. (**a**) Perfect Galaxy International Ltd. EQX0250UV480TN. (**b**) Power-One ULTRA-750-TL-OUTD-4-US (**c**) Power Electronics FS0900CU.

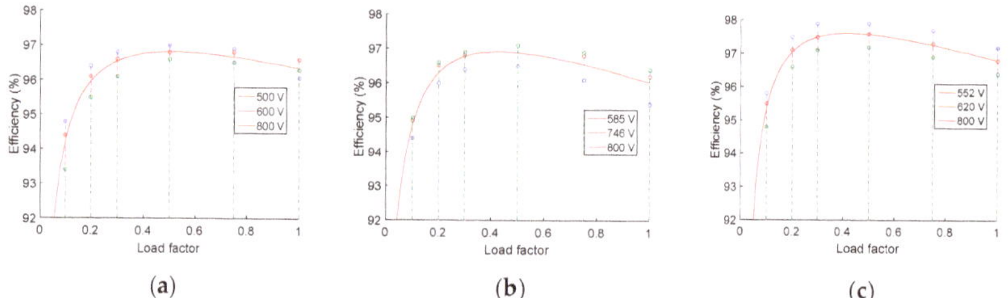

(a)   (b)   (c)

Figure 5. Efficiency curves calculated by means of the Dupont model. (**a**) Perfect Galaxy International Ltd. EQX0250UV480TN. (**b**) Power-One ULTRA-750-TL-OUTD-4-US. (**c**) Power Electronics FS0900CU.

Figure 6a–c show the efficiency surfaces of the inverters, which have been computed by means of the Rampinelli model in the whole range of MPPT voltages. Figure 6d–f detail these results only for the three values of the DC voltage given by CEC. In Figures 7–9, the same results are depicted, obtained in the same conditions, but in these cases by means of Rampinelli nonlinear model, Sandia model, and Driesse model, respectively. As expected, the results obtained by using bidimensional models significantly improve compared to the ones achieved by means of the unidimensional ones. Regarding the dependence of the efficiency curves with v_{in}, note that for the inverter #2 (Power-One ULTRA-750-TL-OUTD-4-US) there are no significant differences between the results offered by the four evaluated bidimensional models. The reason for that is the strong linear dependence with the DC voltage that the efficiency curves of this inverter present. In contrast, in the case of the other two inverters under study, the dependence of the efficiency curves with the DC voltage is not linear and, therefore, the results achieved by means of Rampinelli nonlinear and the Driesse models fit better with the actual CEC data than the Rampinelli and the Sandia models.

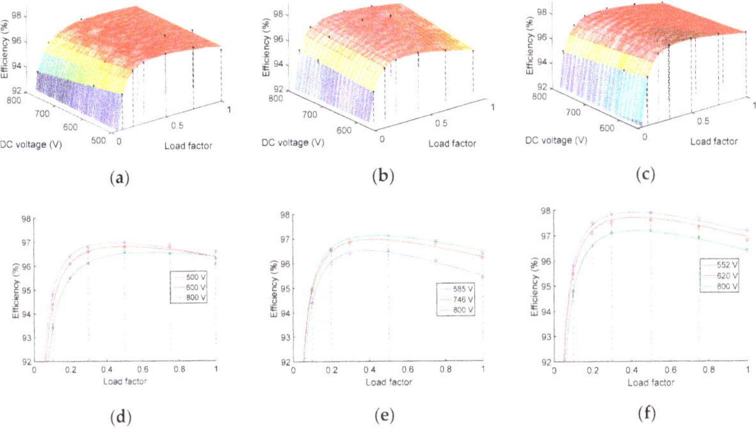

Figure 6. Efficiency surfaces and detail of curves for three values of the DC voltage calculated by means of the Rampinelli model. (**a**,**d**) Perfect Galaxy International Ltd. EQX0250UV480TN. (**b**,**e**) Power-One ULTRA-750-TL-OUTD-4-US. (**c**,**f**) Power Electronics FS0900CU.

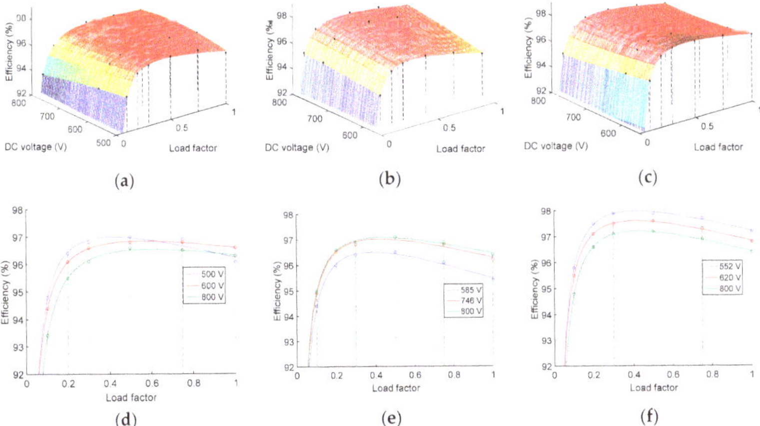

Figure 7. Efficiency surfaces and detail of curves for three values of the DC voltage calculated by means of the Rampinelli nonlinear model. (**a**,**d**) Perfect Galaxy International Ltd. EQX0250UV480TN. (**b**,**e**) Power-One ULTRA-750-TL-OUTD-4-US. (**c**,**f**) Power Electronics FS0900CU.

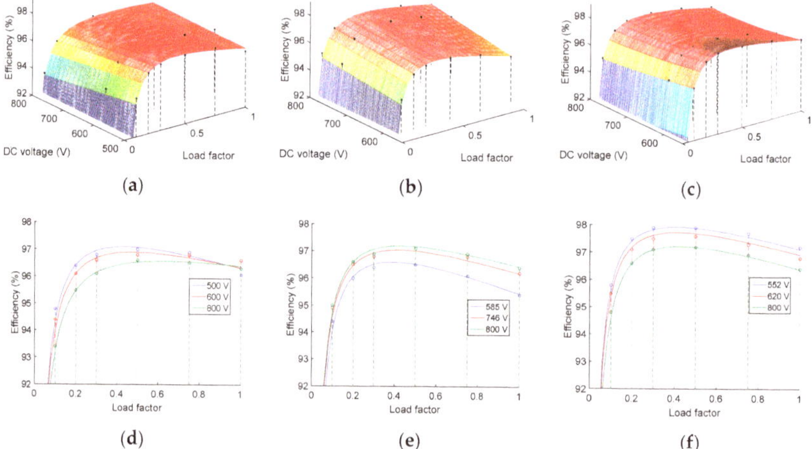

Figure 8. Efficiency surfaces and detail of curves for three values of the DC voltage calculated by means of the Sandia model. (**a,d**) Perfect Galaxy International Ltd. EQX0250UV480TN. (**b,e**) Power-One ULTRA-750-TL-OUTD-4-US. (**c,f**) Power Electronics FS0900CU.

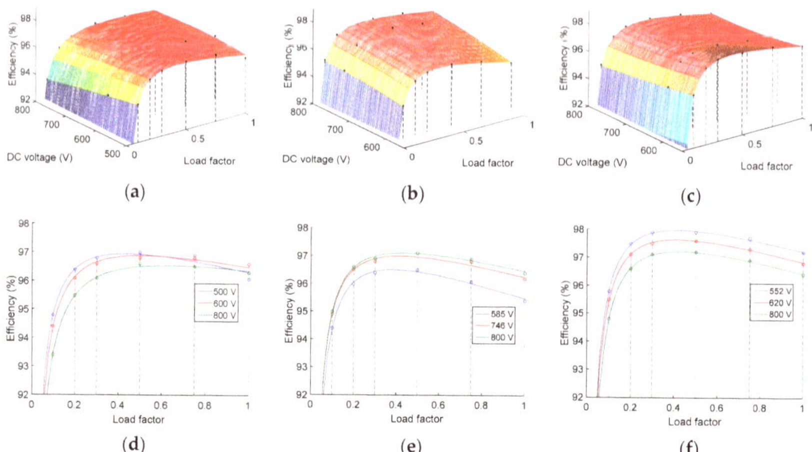

Figure 9. Efficiency surfaces and detail of curves for three values of the DC voltage calculated by means of the Driesse model. (**a,d**) Perfect Galaxy International Ltd. EQX0250UV480TN. (**b,e**) Power-One ULTRA-750-TL-OUTD-4-US. (**c,f**) Power Electronics FS0900CU.

In summary, it may be concluded that the Rampinelli nonlinear and the Driesse models are the best approaches to predict the performance of photovoltaic inverters in terms of efficiency, independent of the relationship between efficiency and MPP voltage of the inverter.

5.2. Evaluation of the Proposed Efficiency-Oriented (EO) Algorithm

The algorithm for the activation/deactivation of the power modules that were described in Section 4 is applied in this section to a central inverter with a nominal power of 3 MW. The algorithm has been tested considering two significant profiles of photovoltaic generation, in sunny and cloudy conditions. Table 11 shows the number of units that are needed to achieve the nominal power with the commercial inverters under study.

Table 11. Number of modules to achieve 3 MW with the inverters under study.

Inverter	Company	Nominal Power	Number of Modules
EQX0250UV480TN	Perfect Galaxy International Ltd.	250 kW	12
ULTRA-750-TL-OUTD-4-US	Power-One	750 kW	4
FS0900CU	Power Electronics	1 MW	3

As explained in Section 3, for a certain set of values of both the load factor and the MPP voltage, the algorithm calculates the optimal number of connected modules to maximize the efficiency of the whole system. To illustrate by means of an example how the algorithm works, Figure 10 shows the optimal number of modules on stream that are calculated by the proposed EO algorithm for a 3 MW central inverter composed by twelve modules of Perfect Galaxy International Ltd. EQX0250UV480TN. The figure depicts the number of inverters in operation that maximizes the efficiency in the whole range of MPP voltages and power.

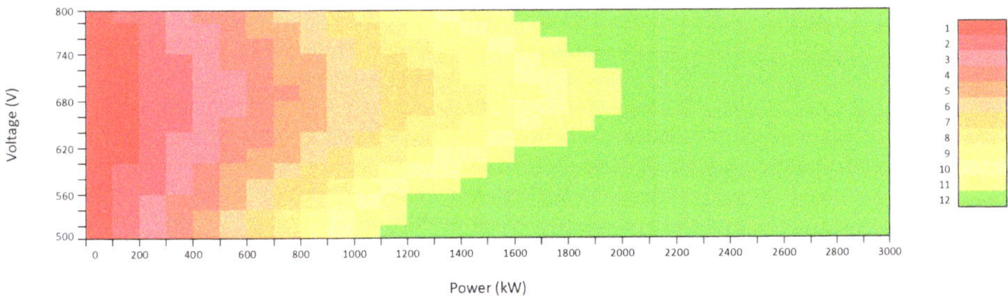

Figure 10. Number of inverters in operation depending on the power generation and the maximum power point tracking (MPPT) voltage.

The proposed algorithm has been implemented in the TMS320F28379D to evaluate the needed computing resources (execution time and memory). To achieve this, a central inverter composed of twelve modules of the Perfect Galaxy International Ltd. EQX0250UV480TN has been considered. Two options for implementing the algorithm have been evaluated. In the first one, the operation map that Figure 10 shows has been programmed by means of a lookup (LU) table. With this approach, the algorithm equations are not solved in real time, so the execution time of the algorithm is expected to be low. In return, the memory requirements increase due to the need of storing in the DSP all the points of the operation map. The second option to implement the algorithm is directly programming the equations in the DSP and solve them in real time. In this case, lower memory requirements and larger execution time are expected than in the case of using a lookup table.

Figure 11a shows the execution time for option #1. In the case under study (12 modules in parallel), the chosen lookup table has a size 15 × 60 (15 input voltages and 60 power levels), needing 1800 bytes of data memory and 47 words of program memory. With this implementation, the algorithm use and its execution time is 540 ns. Note that the memory resources could vary depending on the resolution of the LU table.

Figure 11b shows the measured execution time when the equations of the algorithm are solved in real time. Note that, in this case, the execution time depends on the number of iterations performed by the algorithm, which are related to the number of modules that compose the central inverter and also on the power generated by the PV field. In the case under study, the execution time at low power is 7.54 µs and at high power is reduced to approximately 3.5 µs. The reason for this difference is that at low power, the efficiency must be calculated considering 1 to n inverters on stream. When the power generation increases, the execution time decreases since the algorithm does not calculate the efficiency when the power managed by each module is greater than its nominal power. In other words, the iteration of the loop is not executed when $c_{mod-i} > 1$, as it can be seen in Figure 2. The program memory used in this case is 37 words and the use of data memory is negligible.

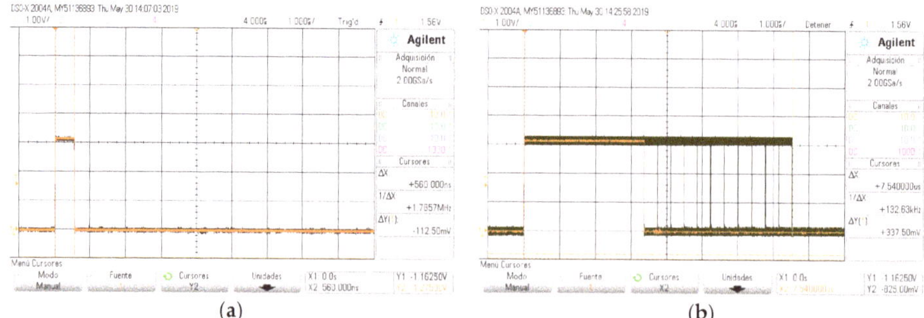

(a) (b)

Figure 11. Execution time of the algorithm (**a**) lookup table implementation. (**b**) equations implemented and solved in real time.

Table 12 summarizes the measured execution times as well as the memory resources for both implementations. The results confirm the expectations about execution time and memory requirements of both kind of implementations, so the choice for a certain application would depend on the need for reducing the implementation time or memory.

Table 12. Execution time and memory resources.

Implementation	Data Memory	Program Memory	Execution Time
Option #1	1800 bytes	47 words	540 ns
Option #2	-	37 words	3.5–7.4 µs

5.2.1. Global Efficiency in the Whole Range of Operation of the PV Farm

Figure 12 depicts the efficiency surfaces obtained by applying the conventional average current-sharing control method (CS) and the efficiency-oriented (EO) method algorithm of activation/deactivation to the 3 MW central inverter described before.

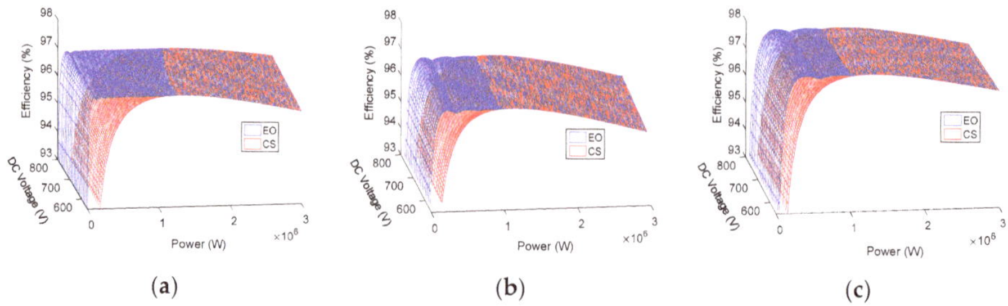

(a) (b) (c)

Figure 12. Efficiency surfaces with average current-sharing control method (**CS**) and efficiency-oriented method (**EO**). (**a**) Perfect Galaxy International Ltd. EQX0250UV480TN. (**b**) Power-One ULTRA-750-TL-OUTD-4-US. (**c**) Power Electronics FS0900CU.

Figure 13a–c depict the detail at 500, 600, and 800 V of the efficiency obtained by both methods applied to the central inverter composed of twelve modules of Perfect Galaxy International Ltd. EQX0250UV480TN. Similarly, Figures 14 and 15 show the results considering the central inverters composed of four modules of Power-One ULTRA-750-TL-OUTD-4-US and three modules of Power Electronics FS0900CU, respectively.

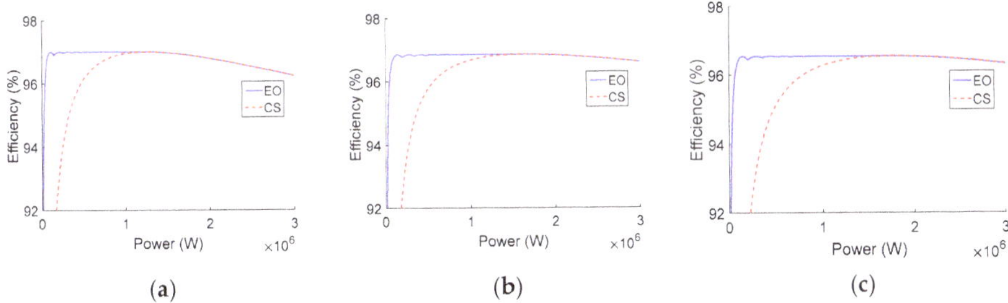

Figure 13. Efficiency of Perfect Galaxy International Ltd. EQX0250UV480TN with average current-sharing control method (**CS**) and efficiency-oriented method (**EO**). (**a**) 500 V. (**b**) 600 V. (**c**) 800 V.

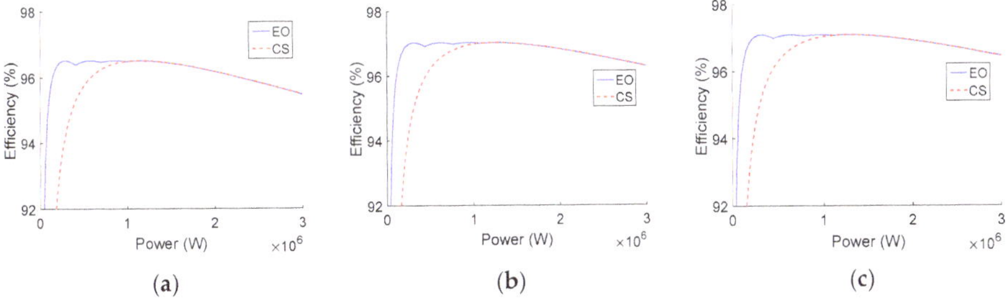

Figure 14. Efficiency of Power-One ULTRA-750-TL-OUTD-4-US with average current-sharing control method (**CS**) and efficiency-oriented method (**EO**). (**a**) 585 V. (**b**) 746 V. (**c**) 800 V.

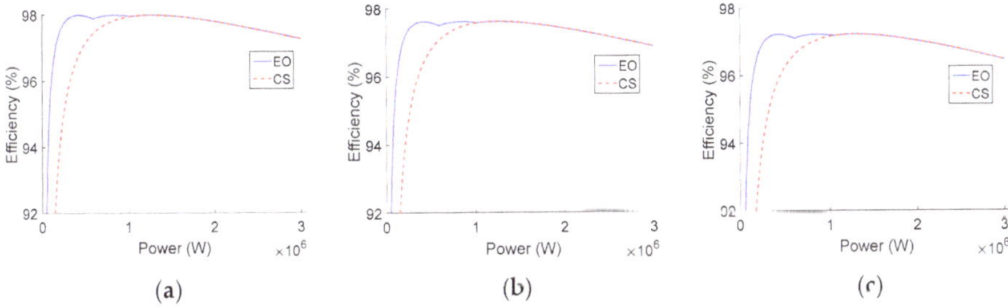

Figure 15. Efficiency of Power Electronics FS0900CU with average current-sharing control method (**CS**) and efficiency-oriented method (**EO**). (**a**) 552 V. (**b**) 620 V. (**c**) 800 V.

These results show that the efficiency-oriented method achieves the best global efficiency in the whole power range independently of the kind of commercial inverter used to build up the central inverter.

5.2.2. Study for a Typical Daily Power Profile

To evaluate the performance of the proposed EO method in realistic conditions, generation profiles in different scenarios have been considered. Figure 16a shows a typical sunny day generation profile while Figure 16b shows a cloudy day generation profile. In the graphics, both the generated power and the DC voltage vary simultaneously.

Figures 17a, 18a and 19a show the efficiency of the evaluated 3 MW central inverter by considering a typical generation profile on a sunny day. Figures 17b, 18b and 19b shows the efficiency of the inverters by considering a typical generation profile on a cloudy day. It can be noticed that, for the central inverters composed by different commercial inverters, the performance of the proposed EO method is clearly better than the efficiency applying the current-sharing method, CS, considering both the sunny and the cloudy day.

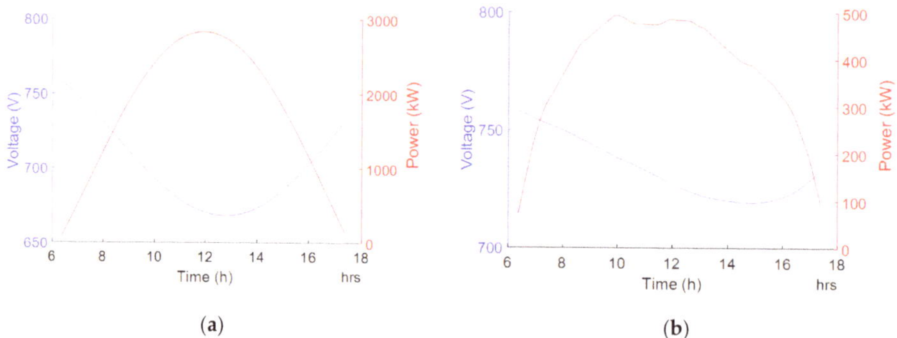

Figure 16. Daily power generation and MPPT voltage curves. (**a**) Sunny day. (**b**) Cloudy day.

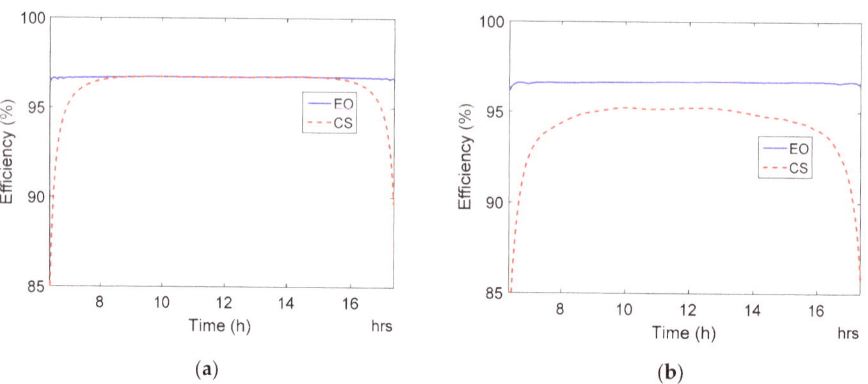

Figure 17. Daily efficiency curves of Perfect Galaxy International Ltd. EQX0250UV480TN with average current-sharing control method (**CS**) and efficiency-oriented method (**EO**). (**a**) Sunny day. (**b**) Cloudy day.

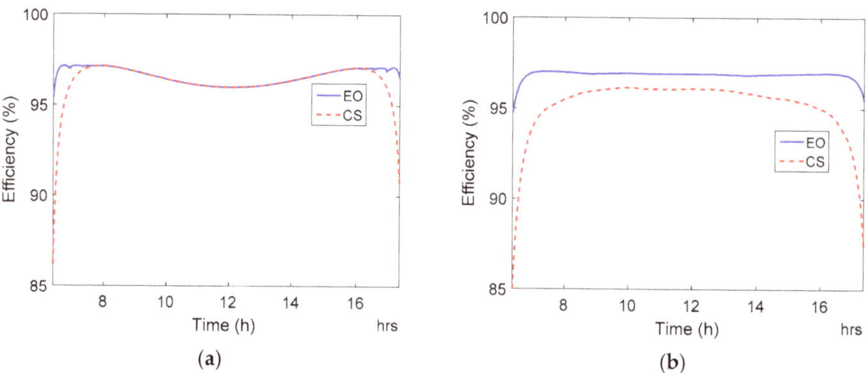

Figure 18. Daily efficiency curves of Power-One ULTRA-750-TL-OUTD-4-US with average current-sharing control method (**CS**) and efficiency-oriented method (**EO**). (**a**) Sunny day. (**b**) Cloudy day.

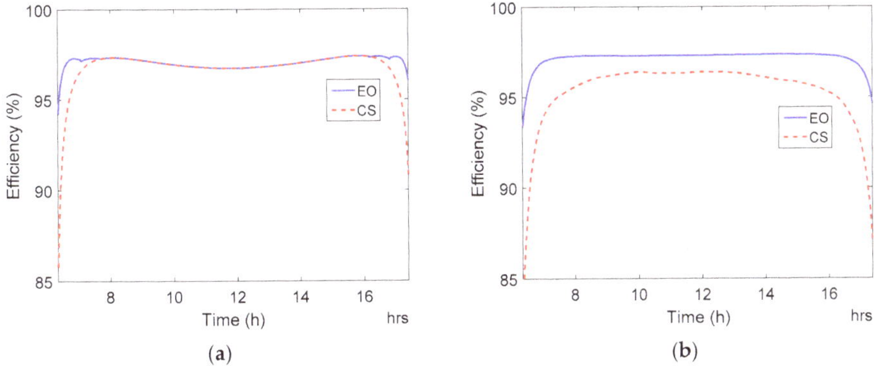

Figure 19. Daily efficiency curves of Power Electronics FS0900CU with average current-sharing control method (**CS**) and efficiency-oriented method (**EO**). (**a**) Sunny day. (**b**) Cloudy day.

6. Conclusions

A control technique to activate/deactivate the power modules of high-power central inverters has been proposed in this paper. The proposed method maintains the advantages of conventional current sharing methods that are usually used to manage the parallel connection of power inverters, as the low need for computational resources, easy implementation, and capability to operate in real time conditions.

The proposed efficiency oriented method is based on a functional model of PV inverters that predicts the efficiency of the system starting from the measurements of the processed power and the MPPT voltage and takes decisions about the number of inverters that should be on stream to improve the global efficiency of the whole central inverter.

A comparative study of several kinds of models to calculate the efficiency of PV inverters has been carried out. The models have been tested by using data of commercial inverters from the "Grid Support Inverters List" published by the California Energy Commission. Unidimensional and bidimensional models have been evaluated, showing that bidimensional and nonlinear models fit much better with the available data. It can then be concluded that bidimensional and nonlinear models are the best choice to be implemented in the proposed EO method.

Regarding the implementation of the proposed algorithm, two options have been evaluated, showing that the execution time can be significantly reduced by implementing the operation map in the whole power and voltage ranges by means of a lookup table. On the contrary, the memory requirements are much lower if the equations of the algorithm are implemented and solved in real time. Therefore, it can be concluded that the choice for a certain application would depend on which factor is preferred to be reduced.

The proposed algorithm of activation/deactivation of the power modules has been applied to a PV field with a nominal power of 3 MW. The inverter's efficiency achieved by the proposed EO method has been compared to the one with the current sharing technique in the whole range of operation of the PV farm. Moreover, the algorithm has been tested considering two significant profiles of photovoltaic generation, in sunny and cloudy conditions. At the beginning and the end of the day in both profiles, when the PV generation is low, the global efficiency is clearly improved with regard to the conventional current sharing method. Moreover, in cloudy conditions, the improvement is significant during all day.

Author Contributions: M.L., E.F. and G.G. proposed the main idea and performed the investigation; M.L. and R.G.-M. developed the software; M.L., E.F., and G.G. wrote the paper. G.G. and E.F. lead the project and acquired the funds for research.

Funding: This work is supported by the Spanish Ministry of Economy and Competitiveness (MINECO), the European Regional Development Fund (ERDF) under Grants ENE2015-64087-C2-2-R and RTI2018-100732-B-C21, and the Spanish Ministry of Education (FPU15/01274).

Conflicts of Interest: The authors declare no conflict of interest. The funders had no role in the design of the study; in the collection, analyses, or interpretation of data; in the writing of the manuscript, or in the decision to publish the results.

References

1. Wu, H.; Locment, F.; Sechilariu, M. Experimental Implementation of a Flexible PV Power Control Mechanism in a DC Microgrid. *Energies* **2019**, *12*, 1233. [CrossRef]
2. Strzalka, A.; Alam, N.; Duminil, E.; Coors, V.; Eicker, U. Large scale integration of photovoltaics in cities. *Appl. Energy* **2012**, *93*, 413–421. [CrossRef]
3. Zhanga, P.; Li, W.; Li, S.; Wang, Y.; Xiao, W. Reliability assessment of photovoltaic power systems: Review of current status and future perspectives. *Appl. Energy* **2013**, *104*, 822–833. [CrossRef]
4. Kim, Y.S.; Kang, S.-M.; Winston, R. Modeling of a concentrating photovoltaic system for optimum land use. *Prog. Photovolt. Res. Appl.* **2013**, *21*, 240–249. [CrossRef]
5. Müller, B.; Hardt, L.; Armbruster, A.; Kiefer, K.; Reise, C. Yield predictions for photovoltaic power plants: Empirical validation, recent advances and remaining uncertainties. *Prog. Photovolt. Res. Appl.* **2016**, *24*, 570–583. [CrossRef]
6. Borrega, M.; Marroyo, L.; González, R.; Balda, J.; Agorreta, J.L. Modeling and Control of a Master–Slave PV Inverter With N-Paralleled Inverters and Three-Phase Three-Limb Inductors. *IEEE Trans. Power Electron.* **2013**, *28*, 2842–2855. [CrossRef]
7. Araujo, S.V.; Zacharias, P.; Mallwitz, R. Highly Efficient Single-Phase Transformerless Inverters for Grid-Connected Photovoltaic Systems. *IEEE Trans. Ind. Electron.* **2010**, *57*, 3118–3128. [CrossRef]
8. Mohd, A.; Ortjohann, E.; Morton, D.; Omari, O. Review of control techniques for inverters parallel operation. *Electr. Power Syst. Res.* **2010**, *80*, 1477–1487. [CrossRef]
9. Su, J.; Liu, C. A Novel Phase-Shedding Control Scheme for Improved Light Load Efficiency of Multiphase Interleaved DC–DC Converters. *IEEE Trans. Power Electron.* **2013**, *28*, 4742–4752. [CrossRef]
10. Ahn, Y.; Jeon, I.; Roh, J. A Multiphase Buck Converter with a Rotating Phase-Shedding Scheme For Efficient Light-Load Control. *IEEE J. Solid-State Circuits* **2014**, *49*, 2673–2683. [CrossRef]
11. Chen, Y.; Hsu, J.; Ang, Y.; Yang, T. A new phase shedding scheme for improved transient behavior of interleaved Boost PFC converters. In Proceedings of the 2014 IEEE Applied Power Electronics Conference and Exposition—APEC 2014, Fort Worth, TX, USA, 16–20 March 2014; pp. 1916–1919.
12. Peng, H.; Anderson, D.I.; Hella, M.M. A 100 MHz Two-Phase Four-Segment DC-DC Converter With Light Load Efficiency Enhancement in 0.18/spl mu/m CMOS. *IEEE Trans. Circuits Syst. Regul. Pap.* **2013**, *60*, 2213–2224. [CrossRef]
13. Costabeber, A.; Mattavelli, P.; Saggini, S. Digital Time-Optimal Phase Shedding in Multiphase Buck Converters. *IEEE Trans. Power Electron.* **2010**, *25*, 2242–2247. [CrossRef]
14. Dupont, F.H.; Zaragoza, J.; Rech, C.; Pinheiro, J.R. A new method to improve the total efficiency of parallel converters. In Proceedings of the 2013 Brazilian Power Electronics Conference, Gramado, Brazil, 27–31 October 2013; pp. 210–215.
15. Huang, W.; Liao, S.; Teng, J.; Hsieh, T.; Lan, B.; Chiang, C. Intelligent control scheme for output efficiency improvement of parallel inverters. In Proceedings of the 2016 IEEE/ACIS 15th International Conference on Computer and Information Science (ICIS), Okayama, Japan, 26–29 June 2016; pp. 1–6.
16. Effler, S.; Halton, M.; Rinne, K. Efficiency-Based Current Distribution Scheme for Scalable Digital Power Converters. *IEEE Trans. Power Electron.* **2011**, *26*, 1261–1269. [CrossRef]
17. Monteiro, L.; Finelli, I.; Quinan, A.; Macêdo, W.N.; Torres, P.; Pinho, J.T.; Nohme, E.; Marciano, B.; Silva, S.R. Implementation and Validation of Energy Conversion Efficiency Inverter Models for Small PV Systems in the North of Brazil. In *Renewable Energy in the Service of Mankind Vol II*; Springer: Cham, Switzerland, 2016; pp. 93–102.
18. Sánchez Reinoso, C.R.; Milon, D.H.; Buitrago, R.H. Simulation of photovoltaic centrals with dynamic shading. *Appl. Energy* **2013**, *103*, 278–289. [CrossRef]

19. Muñoz, J.; Martínez-Moreno, F.; Lorenzo, E. On-site characterisation and energy efficiency of grid-connected PV inverters. *Prog. Photovolt. Res. Appl.* **2011**, *19*, 192–201. [CrossRef]

20. Dupont, F.H.; Zaragoza Bertomeu, J.; Rech, C.; Pinheiro, J.R. A simple control strategy to increase the total efficiency of multi-converter systems. In Proceedings of the Brazilian Power Electronics Conference, Gramado, Brazil, 27–31 October 2013; pp. 284–289.

21. Baumgartner, F.P. Euro Realo Inverter Efficiency: Dc-Voltage Dependency. In Proceedings of the 20th European Photovoltaic Solar Energy Conference, Barcelona, Spain, 6–10 June 2005.

22. Davila-Gomez, L.; Colmenar-Santos, A.; Tawfik, M.; Castro-Gil, M. An accurate model for simulating energetic behavior of PV grid connected inverters. *Simul. Model. Pract. Theory* **2014**, *49*, 57–72. [CrossRef]

23. Rampinelli, G.A.; Krenzinger, A.; Romero, F. Mathematical models for efficiency of inverters used in grid connected photovoltaic systems. *Renew. Sustain. Energy Rev.* **2014**, *34*, 578–587. [CrossRef]

24. King, D.L.; Gonzalez, S.; Galbraith, G.M.; Boyson, W.E. *Performance Model for Grid-Connected Photovoltaic Inverters*; Sandia National Laboratories: Albuquerque, NM, USA, 2007.

25. Driesse, A.; Jain, P.; Harrison, S. Beyond the curves: Modeling the electrical efficiency of photovoltaic inverters. In Proceedings of the 2008 33rd IEEE Photovoltaic Specialists Conference, San Diego, CA, USA, 11–16 May 2008; pp. 1–6.

26. California Energy Commission Grid Support Inverters List. Available online: https://www.gosolarcalifornia.ca.gov (accessed on 14 May 2019).

27. Mathai, A.M.; Haubold, H. *Fractional and Multivariable Calculus: Model Building and Optimization Problems*; Springer: Cham, Switzerland, 2017; ISBN 978-3-319-59993-9.

28. MathWorks Statistics and Machine Learning Toolbox. Available online: https://www.mathworks.com (accessed on 14 May 2019).

Communication

Snapshot of Photovoltaics—February 2019 [†]

Arnulf Jäger-Waldau

European Commission, Joint Research Centre (JRC), Via E. Fermi 2749, I-21027 Ispra (VA), Italy;
Arnulf.jaeger-waldau@ec.europa.eu; Tel.: +39-0332-789119
† The scientific output expressed is based on the current information available to the author, and does not imply
  a policy position of the European Commission.

Received: 8 February 2019; Accepted: 19 February 2019; Published: 26 February 2019

Abstract: Over the last two decades, grid-connected solar photovoltaic (PV) systems have increased from a niche market to one of the leading power generation capacity additions annually. In 2018, over 100 GW of new PV power capacity was added. The annual PV capacity addition in 2018 was more than the total cumulative installed PV capacity installed until the mid of 2012. Total installed PV power capacity was in excess of 500 GW at the end of 2018. Despite a 20% decrease in annual installations, China was, again, the largest market with over 44 GW of annual installations. Decentralized PV electricity generation systems combined with local battery storage have substantially increased as well.

Keywords: renewable energies; photovoltaic (PV); energy challenge; policy options; technological development; market development; battery storage

---

1. Introduction

The urgent need for a de-carbonization of the power sector was stressed again during the 24th session of the Conference of the Parties (COP24) meeting in Katowice, Poland in December 2018. In 2015, the average CO_2 emission per kWh of electricity was about 506 g globally [1]. The World Energy Outlook 2017 New Policy Scenario of the International Energy Agency (IEA) predicts that those emissions should decrease to 325 g CO_2/kWh by 2040. The situation looks somewhat better in Europe. There, the emissions per kWh of electricity should decrease form 344 g CO_2/kWh in 2016 to roughly 150 g CO_2/kWh in 2040. However, such a decrease is still not sufficient for the necessary reduction of CO_2 emissions. To honor the Paris Agreement, a maximum of 65 g CO_2/kWh is allowed [2]. The only scenario that meets this requirement is the Sustainable Development Scenario. Under this scenario the emissions from electricity production in Europe have to decrease to 45 g CO_2/kWh.

The crucial role of solar photovoltaics (PVs) to achieve this goal in a cost-effective manner was outlined in a number of 100% renewable energy source (RES) scenarios. Solar PV power generation has the potential to increase from about 600 TWh (2.4%) in 2018 to 6300 TWh (22%) in 2025, and surpassing 40,000 TWh (up to 70%) in 2050 [3]. To achieve such ambitions, the corresponding PV power capacities have to increase from slightly more than 500 GW at the end of 2018 to more than 4 TW by 2025, and to 21.9 TW by 2050 world-wide. This requires a growth of the annual market from slightly over 100 GW in 2018 to a few hundred GW thereafter.

Over the last two decades, the growth dynamics of PV deployment has changed from government driven incentive programs to market driven investment decisions. Besides the financial aspects and changing political framework conditions, the rapid growth was made possible by thorough technology achievements. Major achievements in material and solar cell research, and progress in manufacturing technology, have made the transition possible. Environmental and health concerns over the use of fossil energy sources, the increased volatility and upward pressure of fossil energy prices, and the commitment of many countries to the Paris Agreement are all adding momentum to it.

2. Photovoltaic Solar Cell Production

Reports of global solar cell production in 2018 vary between 110 GW and 120 GW. As more and more companies go private, manufacturing data collection gets more complicated. In addition, there is no common reporting format. Only a few companies report production figures, whilst others report shipment or sales figures. This explains the considerable uncertainty in this data.

Manufacturing data for this communication were collected through public stock market reports, commercial market reports, as well as through personal contacts. The different data sets were then compared, and this resulted in an estimate of 113 GW produced in 2018 (Figure 1). This corresponds to an annual increase of about 7% compared to 2017.

Production Statistics Uncertainties:

- Solar cell or thin film module production data are reported by a few companies only.
- Products in stock, but produced in the previous year, can be included in shipment data.
- The report of "solar products" in the shipment figures, which is often used, in general does not differentiate between the different products like wafers, cells, or modules.
- The risk of double counting has increased with the major uptake of original equipment manufacturing (OEM).

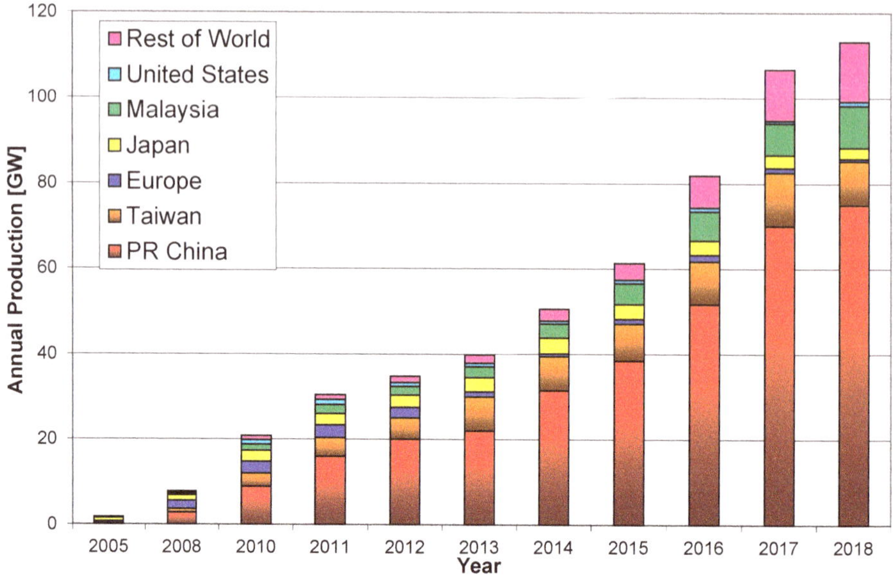

Figure 1. World photovoltaic (PV) cell/module production from 2005 to 2018.

Solar system hardware prices have declined by over 80% over the last two decades. Over the last 10 years the Levelized Cost of Electricity (LCoE) benchmark has decreased by over 75% to USD 69/MWh. These developments were made possible by the rapid increase of countries embracing solar energy, and the rapid growth of the PV manufacturing industry in China after 2005. However, the consequence was not only a massive market growth, but a severe price pressure, which resulted in a major consolidation in the PV manufacturing industry [4]. Despite a significant number of bankruptcies and low profit margins in the manufacturing part of the PV value chain, there is a significant number of new market entrants.

One of the fastest growing companies is Tongwei Solar. The company was set up just six years ago in 2013 as part of the Tongwei Group. The latter is a private company that has its core business in agriculture and new energy. In 2011, Tongwei Group and the Xinjiang Government agreed on an integrated PV strategic cooperation project. This project included the setup of a 50,000 ton solar-grade polysilicon plant, a 3 GW manufacturing capacity for solar wafers and solar cells, and last but not least, 5 solar power plants. In 2018, Tongwei Solar reported an increase of its annual production capacity to 80,000 tons of polysilicon, and 12 GW for solar cells and solar modules. With a polysilicon production of about 17,000 tons and solar cell shipments of 3.85 GW, the company has already ranked 6th for both products in 2017 [4].

Overall, there are still new capacity announcements, which will increase the total manufacturing capacity significantly. The rational for these expansion plans are the expectations of an annual 40 to 50 GW market in China, a continuation of market growth in India, and new markets in Africa, the Middle East, and South America. Further manufacturing cost reductions are expected through Manufacturing 4.0 factories, improved solar cell efficiencies, and reduced material consumption along the whole cell and module manufacturing process. PV manufacturers with older production equipment are suffering the most as they have to compete with new entrants, which have the advantage of a lower manufacturing capital expenditure (CAPEX) and higher efficiency products. For example, CAPEX in 2018 for a polysilicon plant with an annual production capacity of 10,000 tons has decreased by 90%, compared to the USD 1.5 billion in 2006–2007. Electricity represents about 20% to 40% of wafer and polysilicon production costs; therefore, the industry is looking for manufacturing sites with the lowest costs. This can be seen in China. The country's northwestern and southwestern regions have attracted major investments in new manufacturing plants for polysilicon and wafers due to the fact that power prices can be up to 90% lower than in the Eastern coastal regions [5].

The overall number of jobs in solar photovoltaic electricity will significantly increase in line with the expanding PV markets. However, most of the jobs will be downstream in the value chain. The main reason for this development is the increasing implementation of manufacturing 4.0 with a low workforce for solar cell and module manufacturing.

3. Solar PV Electricity generation and Markets

Since 2009, the weighted benchmark levelized costs of electricity (LCOE) for non-tracking crystalline silicon PV systems has decreased from about USD 225/MWh in the first half of 2009 to USD 60/MWh in the second half of 2018 [6,7]. This corresponds to a reduction of almost 75% over the last decade.

Actual electricity generation costs from photovoltaic systems depend on various factors like solar radiation, type of system, fixed operation and maintenance (O&M) costs, as well as finance conditions, which can differ significantly from country to country. In the second half of 2018, the range of LCoE for non-tracking systems varied between USD 38/MWh and USD 147/MWh, and between USD 41/MWh and USD 83/MWh for 1-axis tracking systems [7].

The prices for Power Purchase Agreements (PPA) can be even lower, especially in sunnier regions of the world with low financing costs. An example of the importance of stable and low-cost financing conditions is the result of the PV tender in Senegal, where the projects will be financed under the International Finance Corporation (IFC)-backed Scaling Solar initiative [8]. In April 2018, Senegal's Commission de Régulation du Secteur de l'Electricité (CRSE) announced the tender results to build two 60MW solar PV plants. The winning bids were EUR 38.026/MWh and EUR 39.83/MWh (USD 43.28/MWh and USD 45.33/MWh).

Despite a decrease of investments in solar energy to USD 130.8 billion (−24%), the annual installations modestly increased, by about 10%, to 109 GW in 2018 (Figure 2) [9]. The reasons for the decline were the lower capital costs for solar photovoltaic systems on the one side, and a decline of the PV installations in China by roughly 20%, compared to 2017, on the other side.

Most, but not all, market analysts expect a larger growth rate in 2019. New installations are forecasted to be between 107 GW and 140 GW [9,10]. The IEA's Renewable Energy Market Report 2018 forecasts a new photovoltaic power capacity between 575 and 720 GW that will be installed globally between 2018 and 2023 [11].

Market Statistics Uncertainties:

- Some statistics report system hardware installations, others report the actual connection to the grid or the start of electricity delivery. Missing grid capacities or administrative reasons have a major impact on the later.
- The installation figures in this communication report the physical installation of the PV system.
- Capacity figures can either be reported in nominal DC peak power (Wp) under standard test conditions (1000 W irradiance, air mass 1.5 light spectrum, and 25 °C device temperature) or utility peak AC power. In some statistics, both capacities are sometimes mixed, e.g., Eurostat.
- PV capacity figures in this communication are nominal DC peak power (Wp) for reasons of consistency.
- Not all countries have official PV capacity statistics, system installations, nor sales statistics.

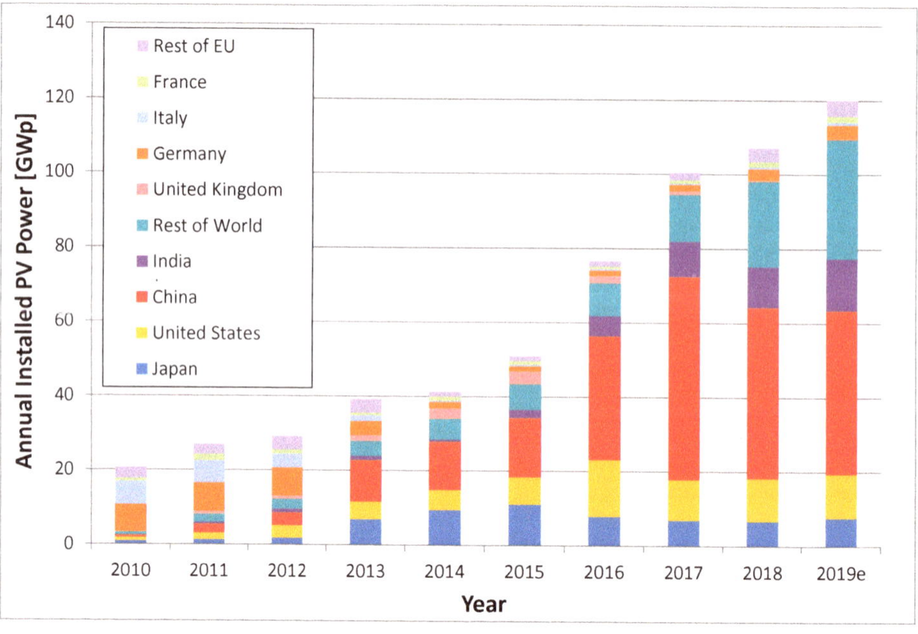

Figure 2. Annual photovoltaic installations from 2010 to 2019 (data source: [12–14], and own analysis).

At the end of 2018, China was home to roughly one-third of global installed PV capacity (about 180 GW). The European Union was second with about 23% (or 117 GW), followed by the United States of America with 12% (or 63 GW) (Figure 3).

Africa: Africa has vast solar resources, and the electricity generation from solar photovoltaic systems can be twice as high in large parts of Africa compared to Central Europe. Despite these advantages, solar photovoltaic electricity generation is still limited. Solar home systems (SHS) or solar lanterns were the main applications until the end of the last decade. The statistics for these applications are extremely imprecise, or even non-existent. Major policy changes have occurred since 2012, and the number of utility-scale PV projects, which are in the planning or realization stages, have increased considerably. In 2018 about 1.6 GW of new PV capacity was installed. The main markets were Egypt (>600 MW), Algeria (>200 MW), and Rwanda (180 MW).

Total African (documented) operational PV power capacity was close to 4.5 GW by the end of 2018. For 2020, the targeted capacity is currently in excess of 10 GW.

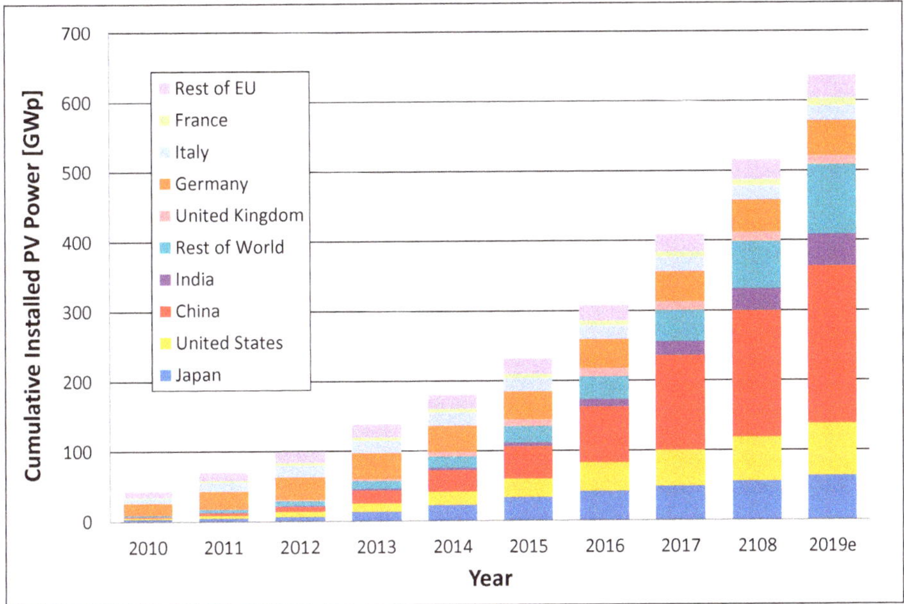

Figure 3. Cumulative photovoltaic installations from 2010 to 2019 (data source: [12–14], and own analysis).

Asia and Pacific Region: Despite the 20% decrease in new photovoltaic electricity system installations in China, the market remained almost stable due to significant market increases in Australia, India, and South Korea, as well as market uptakes in a number of countries in the Middle East and Southeast Asia. With over 44 GW, China was, again, the largest market, followed by India with almost 11.7 GW, Japan with over 6.7 GW, and Australia at 3.8 GW. For 2019, a slight increase to about 80 GW could be possible under stable policy conditions.

European Union: On 14 June 2018, the negotiators from the Commission, the European Parliament, and the Council reached a political agreement regarding the increase of renewable energy use in the European Union [15]. The new, renewable energy target for the EU for 2030 was set at 32%. However, this target is only binding for the EU as a whole, not on Member State levels. The revised renewable energy directive, which included a review clause by 2023 for an upward revision of the EU-level target, was published on 21 December 2018 [16].

After its peak in 2011, when PV installations in the EU accounted for 70% of worldwide installations, six years of market decreases and stagnation followed [17]. This trend was finally reversed when the PV market in the European Union increased almost 50%, from about 6 GW in 2017 to 8.8 GW in 2018. The increase was due to stronger than expected markets in Germany (3.1 GW), the Netherlands (1.4 GW), France (>1 GW), and Hungary (>0.5 GW).

In November 2018, the European Commission published its Vison for 2050, *A Clean Planet for All*, which outlined that the use of renewable energy sources has to exceed 60% by 2050 to reach an average increase of 1.5 °C, or net zero emissions [18]. To meet the EU's new energy and climate targets for 2030, Member States were required to prepare and submit by the end of 2018 a National Energy and Climate Plan (NECP) for the period from 2021 to 2030.

Americas: Markets in North and South America increased by over 25% and added about 17.5 GW of new solar photovoltaic power in 2018. The three largest markets were the USA (11.4 GW), Mexico (2.5 GW), and Brazil (1.5 GW). The number of countries embracing solar photovoltaic energy in Central and South America is increasing, and six countries had a PV market larger than 100 MW in 2018.

4. Conclusions

There is a general consensus amongst investment and energy analysts that solar photovoltaic energy will continue to grow faster than the overall energy demand in the coming years. Various industry associations, as well as Bloomberg New Energy Finance (BNEF), the European Renewable Energy Council (EREC), the Energy Watch Group with Lappeenranta University of Technology (LUT), Greenpeace, and the International Energy Agency have all published scenarios showing the possible growth of PV power capacity [19–23]. In Table 1 the numbers of the different scenario studies are compared. An interesting development can be observed looking at the IEA scenarios, which show a significant increase from the 2016 to 2018 scenario. However, IEA expectations are still at the lower end.

Table 1. Projected evolution scenarios of the world-wide cumulative solar electrical capacities through 2040.

Year	2018 [GW]	2020 [GW]	2025 [GW]	2030 [GW]	2040 [GW]
Actual Installations	516	-	-	-	-
Greenpeace (reference scenario)	-	332	413	494	635
Greenpeace (advanced [r]evolution scenario)	-	844	2000	3725	6678
LUT 100% RES Power Sector	-	1168	3513	6980	13,805
BNEF NEO 2018	-	759	1353	2144	4527
IEA New Policy Scenario 2016	-	481	715	949	1405
IEA 450 ppm Scenario 2016 *	-	517	814	1278	2108
IEA New Policy Scenario 2018 *	-	665	1109	1589	2540
IEA Sustainable Development Scenario 2018 *	-	750	1472	2346	4240

Note: * 2025 value is interpolated, as only 2020 and 2030 values are given.

With forecasted world-wide new installations between 270 and 310 GW in 2019 and 2020, only the 100% RES Power Sector scenario for 2020 is out of reach [9]. Global solar electricity production in 2018 was around 600 TWh and could reach 1000 TWh (or 4%) by 2020.

Funding: This research received no external funding.

Conflicts of Interest: The author declares no conflict of interest.

References

1. International Energy Agency. *CO₂ Emissions from Fuel Combustion*; International Energy Agency: Paris, France, 2017; ISBN 978-92-64-27819-6.

2. International Energy Agency. *World Energy Outlook 2017*; International Energy Agency: Paris, France, 2017; ISBN 978-92-64-28230-8.

3. Breyer, C.; Bogdanov, D.; Aghahosseini, A.; Gulagi, A.; Child, M.; Oyewo, A.S.; Farfan, J.; Kristina Pasi Vainikka, S. Solar photovoltaics demand for the global energy transition in the power sector. *Prog. Photovolt. Res. Appl.* **2017**, 1–19. [CrossRef]

4. Jäger-Waldau, A. *PV Status Report 2018*; Publications Office of the European Union: Luxembourg, 2018; ISBN 978-92-79-97466-3.

5. Hwang, A. Solar Wafer Production to See Over-Capacity by Year-End 2018. *DIGITIMES*, 9 January 2018.

6. New Energy Finance. *Q3 Levelised Cost of Energy Outlook*; EU: Brussels, Belgium, 2009.

7. Bloomberg New Energy Finance. *2H 2018 LCOE Update: PV*; EU: Brussels, Belgium, 2018.

8. Scaling Solar Senegal, Latest Tender Results. 3 April 2018. Available online: http://www.finergreen.com/wp-content/uploads/2017/04/18-04-06-Senegal-Scaling-Solar-Tender-Results.pdf (accessed on 21 February 2019).

9. Bloomberg New Energy Finance. *4Q 2018 Global PV Market Outlook*; EU: Brussels, Belgium, 2018.

10. Solar Power Europe. *Global Market Outlook for Solar Power 2018–2022*; Solar Power Europe: Brussels, Belgium, 2018.

11. International Energy Agency. *Renewable Energy Market Report 2018—Market Trends and Projections to 2023*; International Energy Agency: Paris, France, 2018; 210p., ISBN 978-92-64-30684-4.

12. Solar Power Europe. *Global Market Outlook for Photovoltaics*; Solar Power Europe: Brussels, Belgium, 2018.

13. Masson, G.; Kaizuka, I.; Lindahl, J.; Jaeger-Waldau, A.; Neubourg, G.; Ahm, P.; Donoso, J.; Tilli, F. A Snapshot of Global PV Markets—The Latest Survey Results on PV Markets and Policies from the IEA PVPS Programme in 2017. In Proceedings of the 2018 IEEE 7th World Conference on Photovoltaic Energy Conversion, WCPEC 2018—A Joint Conference of 45th IEEE PVSC, 28th PVSEC and 34th EU PVSEC, Waikaloa, HI, USA, 10–15 June 2018; Article number 8547794. pp. 3825–3828. [CrossRef]

14. EurObserverPhotovoltaic Energy Barometer 2018. In *Systèmes Solaires, le Journal du Photovoltaique*; PE Barometer: London, UK, 2018.

15. European Commission Statement/18/4155. 14 June 2018. Available online: http://europa.eu/rapid/press-release_STATEMENT-18-4155_en.htm (accessed on 30 January 2019).

16. DIRECTIVE (EU) 2018/2001 of the European Parliament and of the Council of 11 December 2018 on the Promotion of the Use of Energy from Renewable Sources (Recast). *Off. J. Eur. Union* **2018**, *328*, 82–209.

17. De Santi, G.; Jäger-Waldau, A.; Taylor, N.; Ossenbrink, H. Realizing solar power's potential in the European Union. *Philos. Trans. R. Soc. A* 2013. [CrossRef] [PubMed]

18. European Commission Communication. *A Clean Planet for all—A European Strategic Long-Term Vision for a Prosperous, Modern, Competitive and Climate Neutral Economy*; European Commission Communication: Brussels, Belgium, 2018.

19. Greenpeace International; European Renewable Energy Council (EREC); Global Wind Energy Council (GWEC). Energy [R]evolution, 5th Edition 2015 World Energy Scenario, October 2015. Available online: http://www.greenpeace.org/international/en/publications/Campaign-reports/Climate-Reports/Energy-Revolution-2015 (accessed on 21 February 2019).

20. Ram, M.; Bogdanov, D.; Aghahosseini, A.; Oyewo, A.S.; Gulagi, A.; Child, M.; Fell, H.-J.; Breyer, C. *Global Energy System Based on 100% Renewable Energy—Power Sector*; Lappeenranta University of Technology: Lappeenranta, Finland, 2017.

21. Bloomberg New Energy Finance. *New Energy Outlook 2018*; EU: Brussels, Belgium, 2018.

22. International Energy Agency. *World Energy Outlook*; International Energy Agency: Paris, France, 2016; ISBN 978-92-64-26495-3.

23. International Energy Agency. *World Energy Outlook*; International Energy Agency: Paris, France, 2018; ISBN 978-92-64-30677-6.

Article

Evaluation of the Impact of High Penetration Levels of PV Power Plants on the Capacity, Frequency and Voltage Stability of Egypt's Unified Grid

Hamdy M. Sultan [2,3], Ahmed A. Zaki Diab [2,*], Oleg N. Kuznetsov [3], Ziad M. Ali [1,4,*] and Omer Abdalla [1,5]

1   College of Engineering at Wadi Addawaser, Prince Sattam bin Abdulaziz University, 11991 Wadi Aldawaser, Saudi Arabia; o.abdalla@psau.edu.sa
2   Electrical Engineering Department, Faculty of Engineering, Minia University, 61111 Minia, Egypt; hamdy.soltan@mu.edu.eg
3   Department of Electrical Power Systems, Moscow Power Engineering Institute "MPEI", 111250 Moscow, Russia; kuznetsovon@mpei.ru
4   Electrical Engineering Department, Aswan Faculty of Engineering, Aswan University, 81542 Aswan, Egypt
5   Electrical Engineering Department, University of Medical Sciences & Technology, 79371 Khartoum, Sudan
*   Correspondence: a.diab@mu.edu.eg (A.A.Z.D.); dr.ziad.elhalwany@aswu.edu.eg (Z.M.A.); Tel.: +2-010-217-77925 (A.A.Z.D.)

Received: 31 December 2018; Accepted: 6 February 2019; Published: 11 February 2019

Abstract: In this paper, the impact of integrating photovoltaic plants (PVPs) with high penetration levels into the national utility grid of Egypt is demonstrated. Load flow analysis is used to examine the grid capacity in the case of integrating the desired PVPs and computer simulations are also used to assess the upgrading of the transmission network to increase its capacity. Furthermore, the impact of increasing the output power generated from PVPs, during normal conditions, on the static voltage stability was explored. During transient conditions of operation (three-phase short circuit and outage of a large generating station), the impact of high penetration levels of PVPs on the voltage and frequency stability has been presented. Professional DIgSILENT PowerFactory simulation package was used for implementation of all simulation studies. The results of frequency stability analysis proved that the national grid could be maintained stable even when the PVPs reached a penetration level up to 3000 MW of the total generation in Egypt. Transmission network upgrading to accommodate up to 3000 MW from the proposed PV power plants by 2025 is suggested. In addition, analysis of voltage stability manifests that the dynamic behavior of the voltage depends remarkably on the short circuit capacity of the grid at the point of integrating the PVPs.

Keywords: photovoltaic; voltage stability; grid capacity; penetration level; frequency stability; Egypt's national grid

---

1. Introduction

Renewable energy sources are recently becoming one of the most promising topics of energy systems and policies in most countries. Among the different renewable sources, photovoltaic generation plants have reached a fast growth in the last decades with capacities ranged from small residential application to large-scale grid-connected commercial projects [1]. At present, among all countries, in USA, China, and Germany medium and large-scale photovoltaic power plants (PVPs) have drawn more attention. With the rapid increase in the penetration level of such renewable sources and dispense with the conventional power plants, power systems are anticipated to face changes in their steady-state and dynamic performance. Consequently, integrating high generation from irregular PVPPs creates supplementary challenges to support the transmission networks stability, not only during normal operation but also in the occurrence of abnormal disturbances [2,3]. These abnormal conditions include different types of faults, to which a bus bar or a transmission line might be subjected (three phase to ground fault, single phase to ground fault, phase to phase fault, and etc.), tripping of a mean transmission line, outage of large conventional generating stations and heavy change in load. Accordingly, the expected operating scenarios should be directed and studied in anticipation in order to maintain the transmission network's stability and the power supply reliability throughout the days and nights.

Egypt is rich in renewable energy resources as it is one of the countries located in the solar belt region, which is most suitable for implementation of solar energy projects. The results of Atlas Egypt, depending on the average values of the last 20 years, show that the solar radiation in average ranges between 2000–3200 kWh/(m^2 year) and the duration daily sunshine hours fluctuates between 9–11 hours/day, which creates good opportunities for investment in different fields of solar energy [4,5]. Global horizontal irradiation (GHI) over the territory of Egypt is presented in Figure 1. According to the annual report of the Egyptian electricity holding company (EEHC) published in June 2016, there is a national plan to provide 20% of the demand for energy from renewable sources by 2027, while in this plan wind energy provide 12%, hydroelectric 5.8% and solar energy 2.2%. The plan anticipates significant participation from the private sector, which planned to reach 67% of the desired generation in the plan of the New and Renewable Energy Authority [6,7]. Renewable Energy sector plans to implement 51.3 GW from nonconventional energy sources to the present installed capacity of the national power system. The first integration of renewable sources of energy was in 2011 from the solar thermal part of the power station built in Kuraymat in the south of Egypt with a share of 20 MW from a total capacity of 140 MW. A 10 MW solar photovoltaic plant has entered service in 2015 in the Siwa Oasis in Western Sahara [4]. At the present time, Egyptian electricity holding company is coordinating to take the administrative steps to complete the agreements, which control the process of purchase of the generated energy from the private sector's projects. The project has a total of capacity of 1000 MW distributed as follow: a wind farm in Suez Golf with an installed capacity of 250 MW, a 200 MW photovoltaic power plant in Aswan (Komombo), and a project in the west of Nile River with a total installed capacity of 550 MW from different renewable sources [5].

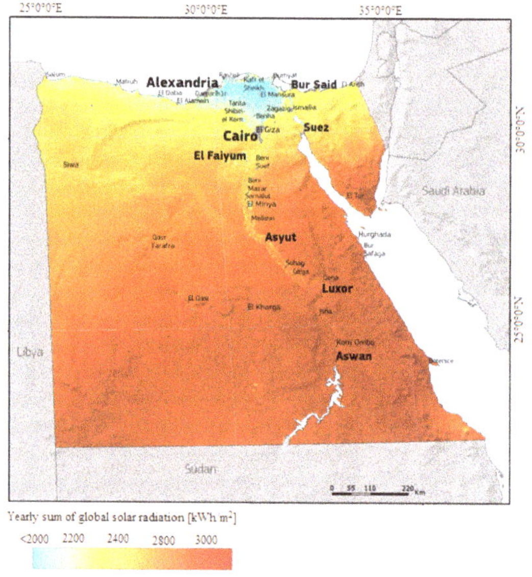

Figure 1. Global horizontal irradiation over Egypt's landscape.

The connection of renewable power plants with high degrees of generation is not a new matter in the most of the developed countries. Taking into account the reports of the International Energy Agency published in 2014 there are about 20 renewable energy power plant in the world, mainly in China and the USA, with a high level of generation of more than 100 MW. An appreciable number of vendors are taking part in the development of the technology that adapts the unusable energy from the sunlight or the wind into usable electrical energy [8]. In view of the above, both the supplier and the operator of the grid have to ensure the connection appropriateness of renewable energy sources to the electrical utility. The National Energy Control Center (NECC) and the Ministry of Electricity and Energy in Egypt issue the technical requirements, defined in the Grid Code that has to be accomplished by any renewable energy projects willing to be integrated into the grid [4,8]. To bypass costly design changes after installation, the performance of these plants is stimulated and examined to meet the applicable technical requirements at the design and pre-design stage. The special simulators are able to model both renewable sources and the electric utility grid, thus inspecting in advance the plant's performance, before the process of on-site testing during commissioning.

The impact of PVPPs with high penetrations was demonstrated and different types of power system stability have been studied [9,10]. In [9] using an equivalent model of Ontario utility grid, Eigen-values and voltage stability have been used in a comprehensive study describing the effect of low and high degrees of photovoltaic generation on the power system stability. While in [10] the impact of PV systems on the short-term stability of voltage has been performed, and the results obtained from this study showed that in the case of voltage sage and the disconnection of PV system, the short-term voltage stability has been strongly impaired. Other authors give attention on the low-voltage ride-through capability of PV power plants with high generation levels [11,12]. In brief, the large PV power plants should remain integrated and into the grid utility in instances of heavy disturbance in grid voltage, as the outage of this large power may further deteriorate voltage recovering throughout and after fault time [13]. The capability to voltage support, because of the integration of PV power plants, has been studied for a wide range of generation [14,15]. Voltage stability denotes to the power system's ability to maintain stable voltages on all buses in the system after deviation from a particular initial operating point. The state of the power system enters the region voltage instability when a disturbance or a sharp increase in the current drawn by the loads results in an unmanageable and continuous drop in the voltage at the buses of the system. Instability of voltage appears in the form of a continuous increase or decrease in the voltages at some buses in the power system [16]. The breakdown of the voltage is usually related to the demand for the reactive power of the load that has not been met due to the lack of reactive power production and transmission. The system is called unstable, if the magnitude of the voltage at one bus in the system, at least, is decreased when increasing reactive power injected into the same point.

Many authors have investigated the study of the impact of high penetration of PV generation on voltage, frequency and power [17–19]. The impact of ambient conditions (Solar radiation and temperature) on the frequency and power of large-scale PV power plant at the point of connection with the grid is conducted in [20]. Studying the techniques of integrating PVPs and WPPs with high power production into the transmission and distribution networks and their impact on the frequency of the power system has been examined [21–23].

The stability analysis of Egypt grid has been discussed in [24–26]. The analysis did not take into account the PV systems and its effect on the grid. In [24], the analysis of the Egypt grid with wind energy plant of Gabl El-Zite wind farm has been discussed. In this reference, the frequency stability did not take in the consideration. In [26], the impact of small PV plants on the stability and performance of the Egypt grid has been introduced. As a result of the complexity and change of the power system structure of Egypt, more analysis are asked to determine the voltage stability, frequency stability and the requirements to interconnect new large-scale PVPs.

Egypt's national grid is hugely extended with new cities, which are established related to the population density and with the new growing industrial areas, which demand more energy.

This extension will be more and more in the future because of the government plan for industrial development. In this paper, study and analysis of the national grid of Egypt connected with planned PVPs will be presented. The Impact of Large PV Plants with respect to the capacity, frequency and voltage stability of Egypt's national utility grid is discussed in details. Egypt's national electric network has been simulated and tested using DIgSILENT software (DIgSILENT, 2017) with the connection of the suggested PVPs.

DIgSILENT has set standards and trends in power system modelling, analysis and simulation for more than 25 years. The proven advantages of the PowerFactory software are its overall functional integration, its applicability to the modelling of generation, transmission, distribution and industrial grids, and the analysis of these grids' interactions. It's considered a software and consulting company providing engineering services in the field of electrical power systems and has a particular interest in the fields of simulation and grid integration of renewable energies. Also, it will allow to for example automatically identify the over/under and loading elements in the power grid, also it can help to identifies exactly the suitable bus that can carry the new load. [27–38]. Moreover, DIgSILENT PowerFactory offers a range of load flow calculation methods, including a full AC Newton-Raphson technique (balanced and unbalanced) and a linear DC method. The enhanced non-decoupled Newton-Raphson solution technique with current or power mismatch iterations, typically yields round-off errors below 1kVA for all buses. The implemented algorithms exhibit excellent stability and convergence. Several iteration levels guarantee convergence under all conditions, with optional automatic relaxation and modification of constraints. Many authors use the DIgSILENT as a benchmark to simulate and to analyze the power system load flow problem [27–31], Newton-Raphson Load flow method modeling by using DIgSILENT is explained in [27,28]. For load flow study and grid simulation, the DIgSILENT is preferred and is recommended in a comparison between many software packages that is because it behaved as it is expected [27–37].

This paper is considered as a part of a project studying the influence of the increasing the level of renewable generation from PV and wind turbines. The first step in achieving this work was collecting the real data of the component of high voltage (220 kV) and extra high voltage (500 kV) transmission networks, the data of transformer substation and loads. These data have been obtained from the national energy control center and the annual reports of the electricity holding company. Then, the data were organized and analyzed. Depending on the parameters of system components, a complete model of the national utility grid of Egypt have been developed using DigSILENT PowerFactory platform. After that, the according to the plan of the Ministry of Electricity and Energy and our previous study for selecting the suitable locations of large-scale PV power plants over the territory of the country, four sites of installing PV systems are proposed. The capacities of the proposed PV power plants have been determined by examining the grid capacity at the points of connection with the national grid. The main part of this paper is to study the impact of the PV plants on the performance of the national utility grid. The validated model has been used to study and analyze the impact of the planned PV power plants on the voltage and frequency stability of the national power system. The future work is to consider the impact of the wind energy power plants. Moreover, the combination with facts devices to improve the overall stability of the system will be considered in the research plane.

2. National Electric Grid of Egypt

In the last ten years, the Egyptian energy system has witnessed rapid developments. New power stations have been built and extensions in the transmission network were implemented to provide electric energy to the existing loading centers as well as access to most isolated systems. A model for the electric utility grid in Egypt, which may be suitable for academic as well as research purposes, has been explored in our previous work [39]. The starting point towards achieving this objective was the electric map of the unified energy system in Egypt, published on the official website of the Ministry of Electricity and Energy (see Figure 2) [7]. Egypt is electrically connected through 500 kV and 220 kV transmission networks, which extended along the Nile River from Aswan in the south to Alexandria in the far north.

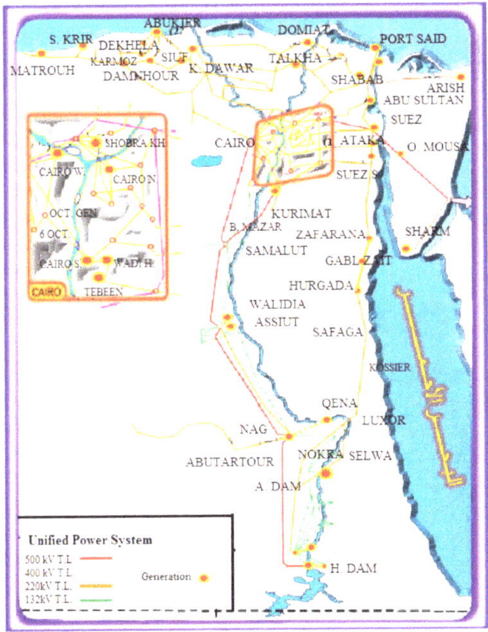

Figure 2. Electricity map of the unified power system in Egypt.

The national utility grid under study is modeled, simulated and assessed using DIgSILENT PowerFactory simulation software package [38]. The following elements were included in the model:

- Transmission lines (500 kV and 220 kV),
- 500/220 kV substations,
- 500/132 kV substations,
- 220/66 kV substations,
- Proposed renewable energy sources and existing conventional plants,
- Electrical connection between Egypt and Jordan on the east side,
- Electrical connection between Egypt and Libya on the west side.

The demands for electric energy are covered, nowadays, from 38 main conventional steam power plants and at least ten of these stations were built more than 35 years ago, two hydroelectric power plants in the south (High-dam and Aswan-dam), in addition to two wind farms on the shores of Red Sea in Zaafrana and Gabl El-Zait. During modeling, the values of the resistances, inductances, capacitances, thermal limits for 500 kV and 220 kV transmission lines and data of the generating stations were obtained from NECC. The parameters of the transformers are taken in accordance with the power limits of the transmission lines. The 220 kV network is divided into six electric zones, similar to the current reality of power transmission companies in Egypt. Single line diagrams for all these regions in addition to the 500 kV grid and technical information about generating stations, substations, transmission lines and nature of loads have been discussed in details in our previous study [39]. The full model includes 218 synchronous machines, 443 transmission lines, 205 substations, 426 bus-bars, 248 transformers and 369 loads. The capacity of the generating stations which are considered in this study is presented in Table 1. Moreover, the data specification of the substations has been listed in Table 2. As stated in the standard of electric energy transmission in Egypt the values of acceptable voltages for different voltage levels under different operating conditions are shown in Table 3 [40,41]. The single line diagram of the power system under study is shown in Figure 3. For conventional power plants, automatic voltage regulators, turbine governors, and power system stabilizers are also included in the model. The existing library of the simulation software has been used for the representation of wind turbine machines and PVPs.

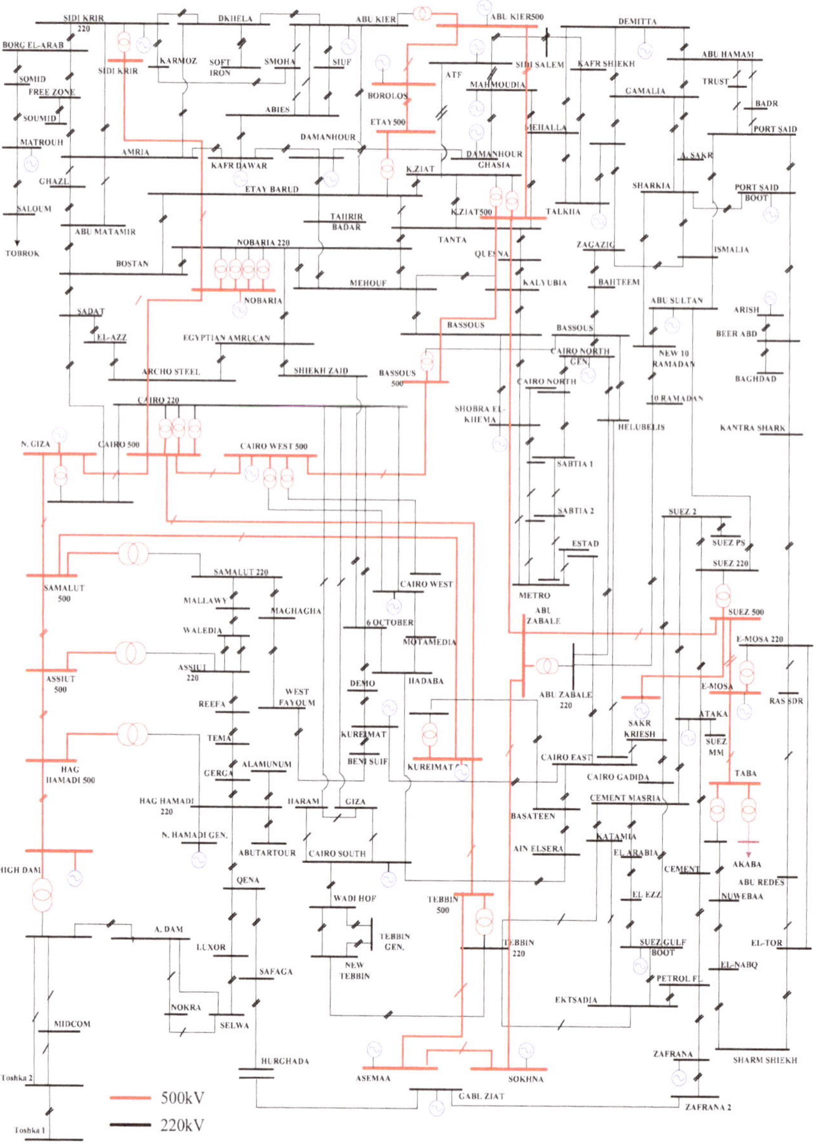

Figure 3. Single line diagram describing the existing transmission network in Egypt.

Table 1. Capacity of the generating stations.

Power Plant	No. of Units	Rating of Each Unit (MW)	Bus Number	Capacity (MW)
HIGH DAM	12	175	1	2100
BENI SUIF	6	400	20	2400
KURAIMAT 1	2	625	10	1250
ASEMAA	2	400	23	800
SOKHNA	2	650	24	1300
SUEZ TH.	1	650	25	650
A.DAM	7	40	586	550
	4	67.5		
CAIRO W.	2	330	6	1360
	2	350		
GIZA N.	9	250	22	2250
NOBARIA	9	250	130	2250
BOROLOS	4	400	28	1600
ABU KIER	2	650	455	1300
T. OCT.	4	150	5031	600
GEN. OCT.	4	150	461	600
CAIRO S.	4	110	518	495
	1	55		
MAHMOUDIA	8	21	566	268
	2	50		
K. DAWAR	4	110	569	440
SHOBRA	4	315	541	1295
	1	35		
SHABAB	3	33.5	300	100.5
ABU SULTAN	4	150	547	600
DOMIAT	6	132	564	1200
	3	136		
TALKHA	8	19.5	559	1406
	3	250		
	2	210		
	2	40		

Table 2. Data specification of the substations.

Substation	Bus Number	Voltage Level (kV)
NOKRA	609	220
A.DAM	586	220
H. DAM	1	500
	490	220
HURGADA	8064	220
NAG HAMADI	501	220
LUXOR	507	220
SAFAGA	608	220
SAMALUT	4	500
	612	220
MALAWI	611	220
MAGAGA	517	220
6 OCTOBER	513	220
N. OCTOBER	388	220
HADABA	512	220
KURAIMAT1	10	550
	515	220
SOKHNA	24	500
GIZA N.	22	500
SELWA	508	220
MEDECOM	316	220
QENA	506	220
CAIRO220	509	220
ZAID	594	220
ASEMAA	23	500
DEKHELA	5020	220

Table 3. The Allowed values of voltage for the transmission networks.

Voltage Level (kV)	Voltage in Case of Normal Operation		Voltage in Case of Contingency Disconnect	
	Higher Voltage	**Least Voltage**	**Higher Voltage**	**Least Voltage**
500	525	475	550	450
400	420	380	420	360
220	231	209	242	198

3. Simulation Results

The technical potential of 27 sites covering all the territory of the country for installation of large-scale grid-connected PV power plants is assessed using software package RETScreen Expert in our previous work [42]. According to this study, the best sites for installation are located in the south of the country along the Red Sea and in the west to the Nile River. Four sites have been suggested for implementation of the new PVPs in KOMOMBO, KOSSIER, MINIA (B. Mazar) and 6.OCT_PV. For integrating the proposed plants with the existing transmission system, four 220 kV double circuit transmission lines are added as follow: 90 km transmission line from KOMOBO to substation A.DAM with a thermal capacity of 850 MW, 70 km from KOSSIER to substation SAFAGA with a thermal capacity of 500 MW, 15 km from MINIA to substation SAMLUT with a thermal capacity of 1200 MW and 10 km from 6.OCT_PV to substation 6.OCT with a thermal capacity of 1500 MW.

3.1. Grid Capacity Assesment

The Newton Raphson load flow method (PowerFactory software) was used to assess the electric network capacity to integrate PVPs in the case of the steady-state operation of the power system and determine the voltage profile and transmission line's power limits.

The capacity of the south Egypt region represented by the 850 MW PV power plant in KOMOMBO is restricted by the thermal limits of the 220 kV lines linking A.DAM to NOKRA and SELWA substations. Moreover, the capacity of the Canal region represented by the 400 MW PV power plant suggested in KOSSIER is restricted by the thermal limits of the 220 kV line between SAFAGA and HURGADA Substations. The capacity of the third site in B. Mazar (MINIA) of 1200 MW to a certain degree bounded by the overloading of the 500 kV line connecting SAMLUT500 and NORTH.GIZA substations. The capacity of the fourth location of 6.OCTOBER City is comparatively high due to the increase in demand for electric energy in that area, so 1500 MW PV power plant suggested in that location. The capacity of the new PV power plants can be increased to twice its suggested values with installing a third 220 kV line between SAFAGA and HURGHADA, 220 kV line between A.DAM and SELWA substations and another 500 kV line between SAMALUT500 and GIZA N. substations.

3.2. Static Voltage Stability Analysis

Voltage and frequency stability has recently become the two important parameters of electric power quality describing the power system performance. Equally important is to know how the elements of the energy system that can stimulate instability work. The voltage is one of the parameters, which has different values at each node in the power system. The voltage at the nodes of the system relies on the values of the impedance of the different elements, which are in the grid or out of the system such as control and protection devices. For maintaining the imposed voltage level at the node, which has a clear impact on the voltage at other nodes in the same zone, different types of voltage adjusting devices can be used. Electric power specialists concern the techniques used for adjusting the voltage level in the power system. The stability of the voltage within the system nodes is very important in the coordination and operation of the system. Problems of voltage instability led to the blackout of energy systems in countries such as Japan in July 1987, in August 2003 in the United States and Canada, in September 23rd, 2003 in southern Sweden and eastern Denmark, and a few days later in Italy and Central Europe as a result of the cascaded outage [43].

In the present study, the stability of voltage of the electric utility grid in Egypt is examined by elaborating the behavior of the power system as seen from four proposed locations; namely A.DAM, SAFAGA, SAMALUT, and 6.OCT. As they are considered the most prospected locations for integration of large photovoltaic generations.

The equivalent short circuit capacity (SCC) has been analyzed. Moreover, the basic definitions of short-circuit conditions are given in the IEC Standard 909 [44]. This standard is based on the calculation of symmetrical initial short circuit current (I"sc), for unloaded networks, i.e., in the absence of passive loads and any shunt capacitance. In order to calculate I"sc, the Thévenin's Theorem is applied to the unloaded network with a source voltage equal to Vn (Vn being the nominal voltage). IEC specifies two standard values for the factor c. The «maximum value» is to be used for apparatus rating purposes and it is fixed at 1.1 for HV systems. The «minimum value» is to be used for other purposes such as the control of motor starting conditions [44], which is typical of fast voltage fluctuations problems such as flicker, and it is fixed at 1 for HV systems. The (IEC standard) short-circuit power is then defined as:

$$S''_{sc} = \sqrt{3} \times V_n \times I''_{sc} \tag{1}$$

The IEC approach perfectly suits either for equipment rating purposes or for non-critical voltage fluctuations problems.

Table 4 presents the equivalent short circuit capacity (SCC), short circuit current and X/R ratio at the four studied locations depending on the short circuit analysis performed used the built-in tools in DigSILENT. The voltage is expected to be more responsive to changes in generated power at sites with less SCC or with higher network resistance.

Table 4. The equivalent of the Grid as seen from the four locations under study.

Location	A.DAM	SAFAGA	SAMALUT	6.OCT
Short Circuit MVA	11305.2	1849.6	14590.9	20838
Short Circuit Current (kA)	29.66	4.854	38.291	54.686
X/R	26.04	7.83	22.12	8.92

Figure 1 presents the grid equivalent circuit as seen from the node of the proposed sites; A.DAM (Site-1), SAFAGA (Site-2), SAMALUT (Site-3) and 6.OCT (Site-4), where PV power plants are proposed to be installed.

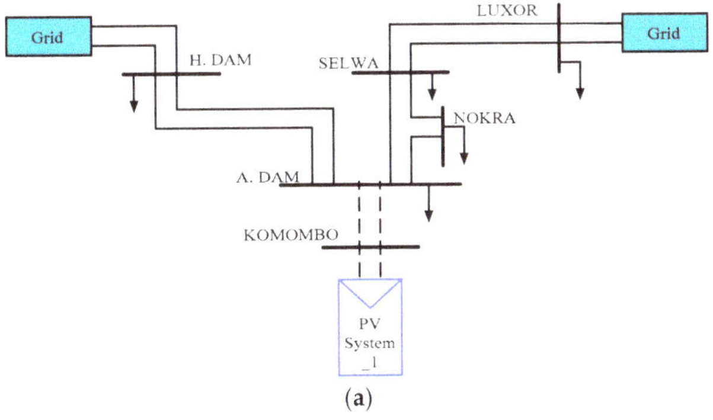

(a)

Figure 4. *Cont.*

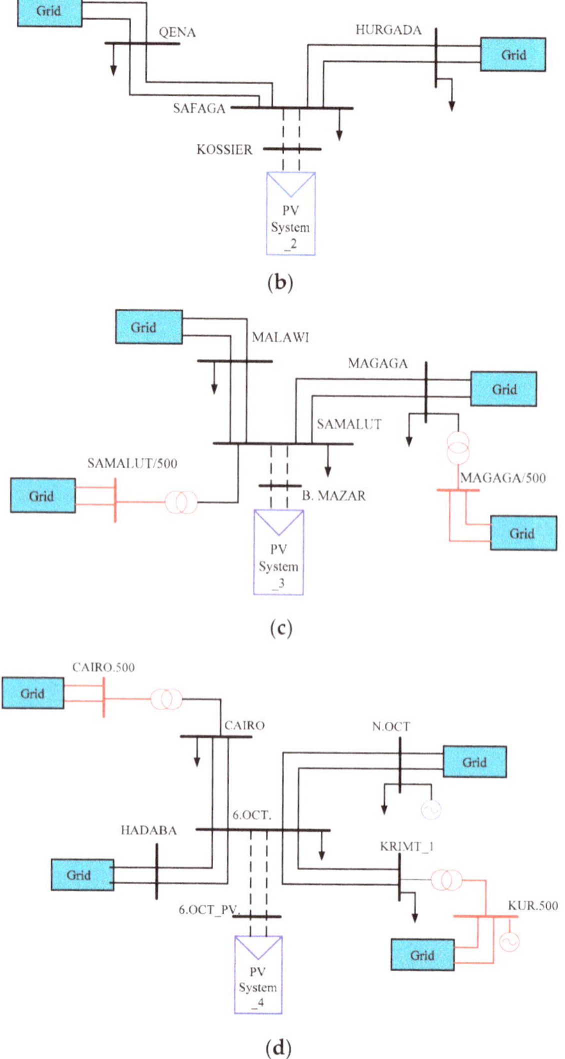

(b)

(c)

(d)

Figure 4. Single line diagram of the equivalent grid as seen from: (**a**) A.DAM (Site-1); (**b**) SAFAGA (Site-2); (**c**) SAMALUT (Site-3); (**d**) 6.OCT (Site-4).

Large-scale photovoltaic power plants would have a particular impact on the voltage stability of the power system integrated into it [15,21]. In this section, the voltage stability limit of PV power plants at different sites is assessed during steady-state normal operation. All the previously mentioned cases that have been simulated are examined to obtain the impact of integration of PVPs with high penetration levels on voltage stability of the Egyptian unified power system. During simulation and studying the behavior of the P-V curves the real power generated at the buses, to which the proposed plants are connected, was gradually increased until voltage collapse is reached.

The Q-V curve is a powerful tool for analyzing the limits of steady-state voltage stability and the network's reactive power margin by illustrating the relationship between the voltage at a certain bus and the reactive power injected to the same bus [45,46]. It demonstrates the distance in reactive power scale from the point of normal operation to the point, at which voltage collapse occurs. The system is called unstable, if the magnitude of the voltage at one bus in the system, at least, is decreased when increasing reactive power injected into the same point. This means that if the sensitivity of V-Q is positive for all buses the system is stable and when the sensitivity of V-Q is negative for one bus, at least, the system is unstable. The driving force for voltage instability is usually the loss loads in a certain section or tripping of transmission lines and other elements by their protective systems leading to cascading outages. Because of these outages, if the generator field current reaches its limit, some generators may lose synchronism [45].

Fundamental to any analysis of the electric power system is the know-how of per unit systems (p.u.). This system is widely used to represent voltages, currents, and impedances in a power system. The per unit systems allows the electrical engineers to solve a single-phase network where: all active power (P) and reactive power (Q) are three phase, voltage magnitudes are represented as a fraction of their original values "base value", all phase angles are presented with their original units.

Per unit (p.u.) system has many advantages over using the standard SI units such as:

- When the electric quantities are expressed in p.u., the comparison with their normal values is straightforward.
- Whatever the power and voltage rating, the values of the impedances in p.u. stay constant.
- Using the p.u. system simplifies the calculation, especially in multi-voltage power systems.

For a given quantity (voltage, current, power, impedance, etc.) the per-unit value is the value related to a base quantity:

$$\text{p.u.} = \frac{\text{quantity expressed in SI system}}{\text{base value}} \tag{2}$$

Generally the base power "S_{Base}" and the base voltage according to the line-to-line voltage "VBase" are chosen, then the value of the base current "I_{Base}" and the base impedance "Z_{Base}" are calculated:

$$I_{Base} = \frac{|S_{Base}|}{\sqrt{3}|V_{Base}|} \tag{3}$$

$$Z_{Base} = \frac{|V_{Base}|^2}{|S_{Base}|} \tag{4}$$

Figures 5–8 show the P-V and Q-V curves at the buses of the four proposed locations. The voltage profiles of the buses, to which PVPs are integrated and the buses of the nearby substations are also reproduced.

From Figures 5–8, it can be noticed that the voltage control has contributed in increasing the penetration level of the PVP at the points of integration by controlling the reactive power injected at these buses. The voltage control method supported the voltage profile at the terminals of the PVPs power stations and the surrounding substations as illustrated in the figures. Moreover, the voltage stability of a certain substation in the grid is directly affected by the equivalent impedance of the electric network as seen from that point. The higher is the equivalent impedance, the higher is the sensitivity of voltage and the lower is the level of real power that can be injected at that point.

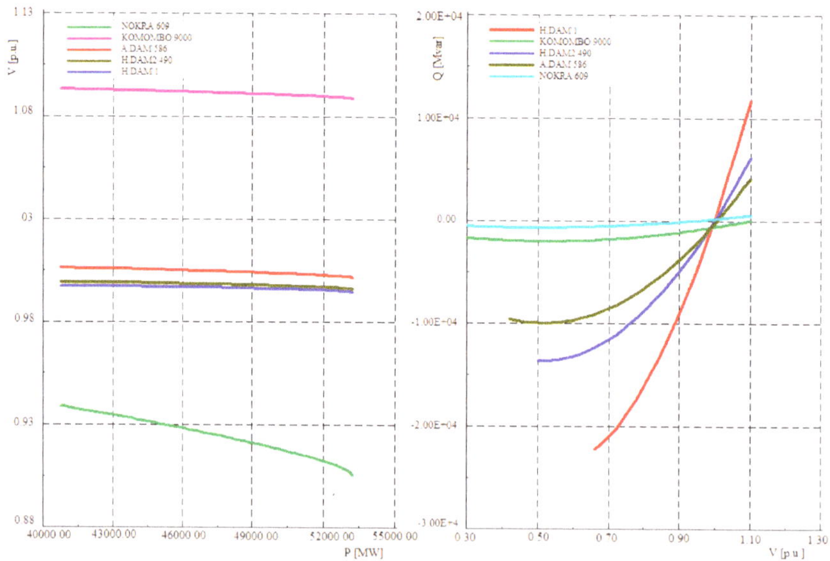

Figure 5. P-V and Q-V Curves of Site-1 (KOMOMBO) and surrounding substations.

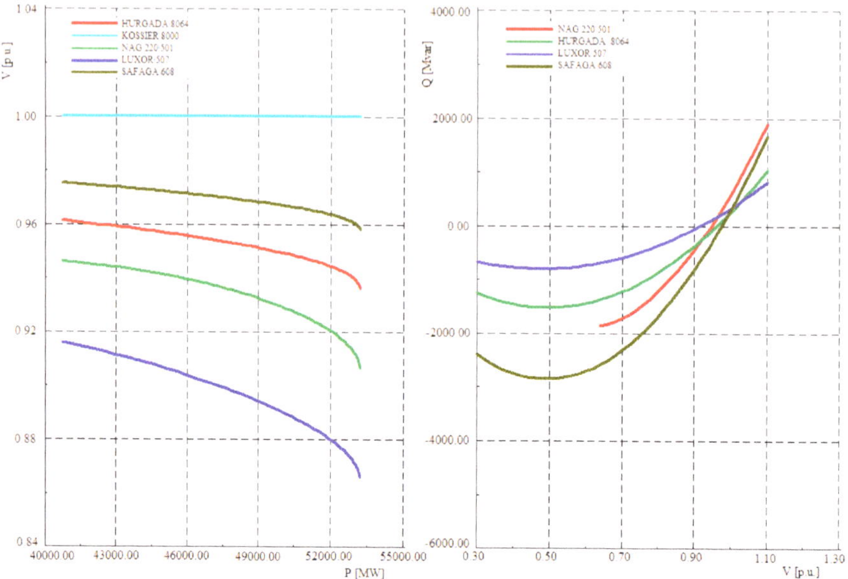

Figure 6. P-V and Q-V Curves of Site-2 (KOSSIER) and surrounding substations.

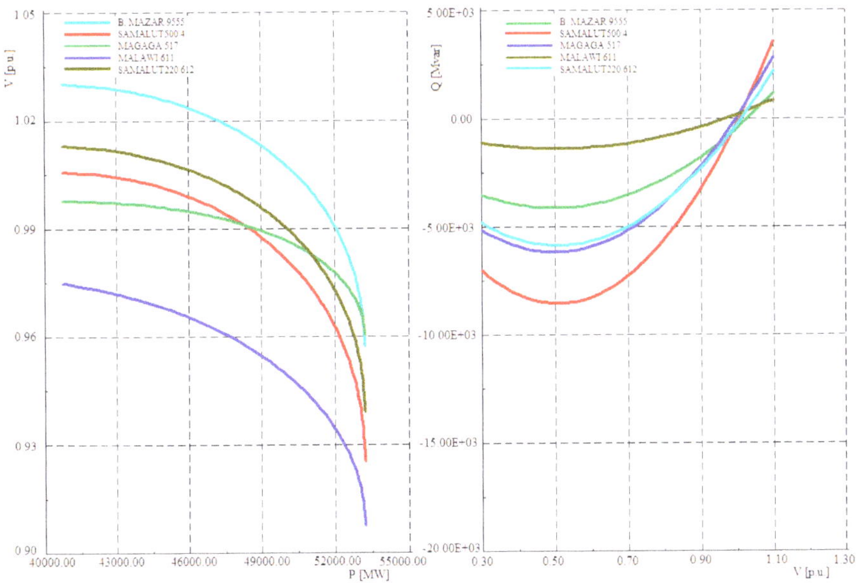

Figure 7. P-V and Q-V Curves of Site-3 (B. MAZAR) and surrounding substations.

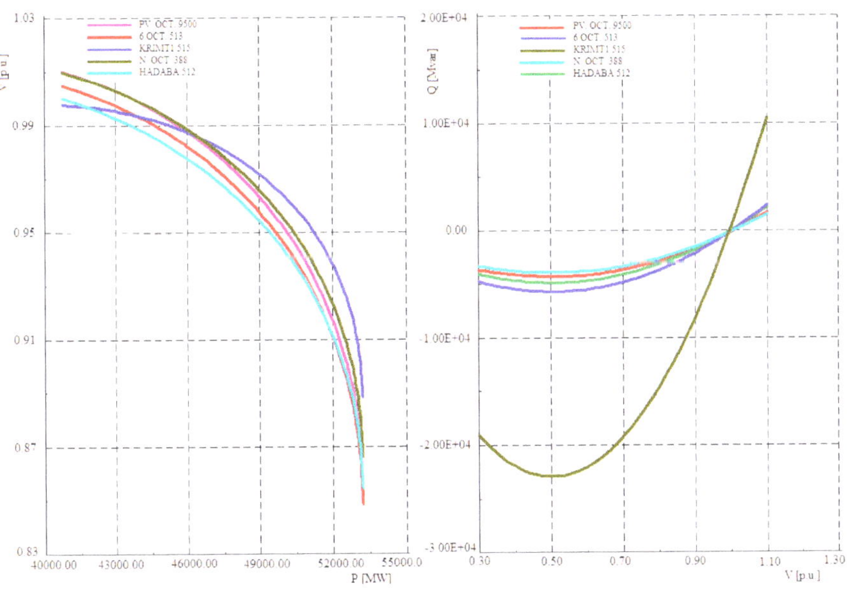

Figure 8. P-V and Q-V Curves of Site-4 (PV OCT.) and surrounding substations.

3.3. Dynamic Voltage Stability Analysis

The dynamic voltage stability of the studied power system was examined by assessing the behavior of the voltage at the buses and the response of the proposed PVPs during the abnormal conditions of three phase to ground fault at the terminals of the proposed PVPs and loss of a generation of a large-scale conventional power plant. The behavior of the system under a three-phase short circuit has been studied and simulated at each of the four proposed sites. The variations in the voltage profile at the buses under fault and the surrounding substations, as well as the active and reactive components of current injected from the suggested PV systems, are established.

3.3.1. Three Phase Short Circuit on the Terminals of PV System_1 (KOMOMBO)

Figure 9 shows the variation in the voltage profile during the application of a 3-phase to ground fault for 0.1 s at the terminals of the high voltage side of transformer substation of PV System_1 (KOMOMBO). The upper graph displays terminal voltage variation at the terminals of the KOMOMBO site. The lower graph represents the variations in the terminal voltage at nearby substations A.DAM, SELWA, NOKRA, LUXOR, which are presented in the single line diagram of Figure 4a and the main power plants. Likewise, Figure 10 presents the variations of the active and reactive components of currents injected from KOMOMBO site and the H.DAM hydroelectric power plant and A.DAM. The results indicate that the system is stable with respect to the same voltage, reactive and active power after fault clearance.

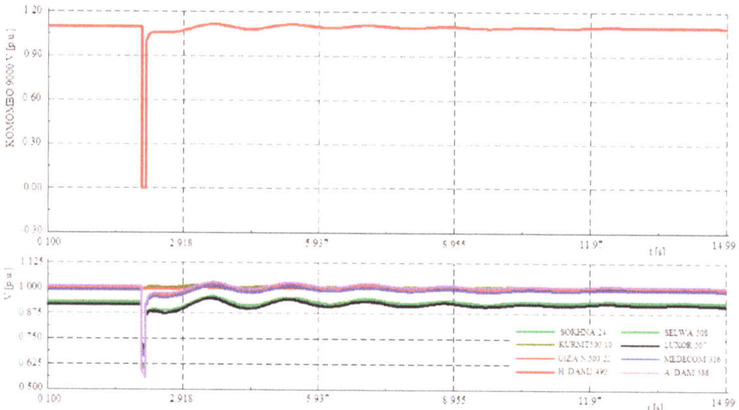

Figure 9. Voltage variation in the case of a 3-phase fault at terminals of PV system at site_1 and nearby bus-bars; Upper: Terminal voltage at PV terminals. Lower: Terminal voltage at nearby buses.

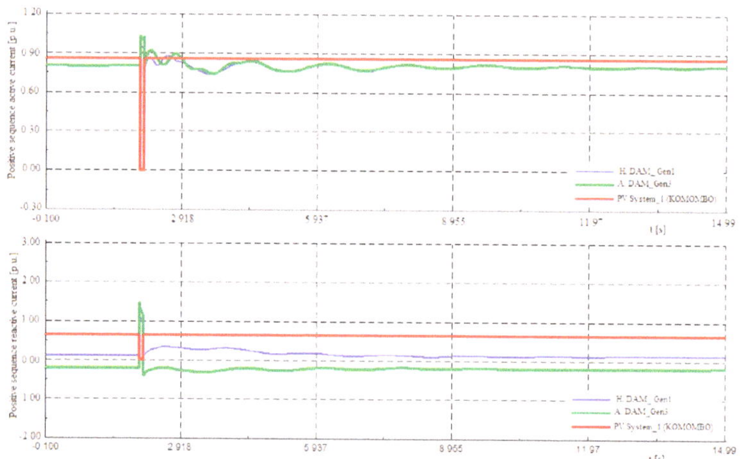

Figure 10. Variations in active and reactive components of current during a 3-phase fault at KOMOMBO (PV System_1) and nearby substations.

3.3.2. Three Phase Short Circuit on the Terminals of PV System_2 (KOSSIER)

Figure 11 shows the variation in the voltage profile in case of the application of a 3-phase fault for 0.1 s at KOSSIER site. The upper graph presents the variation in terminal voltage at the terminals of the KOSSIER power plant. The lower graph shows the terminal voltage variations at surrounding substations mentioned in the single line diagram presented in Figure 4b. Likewise, Figure 12 represents the variations of the active and reactive components of currents injected at KOSSIER site and A.DAM hydroelectric substation. From the results, the system takes more time to reach the stability after fault clearing at the A.DAM hydroelectric substation, because this substation is near the fault location. However, the system each to the stability and has a good dynamic performance.

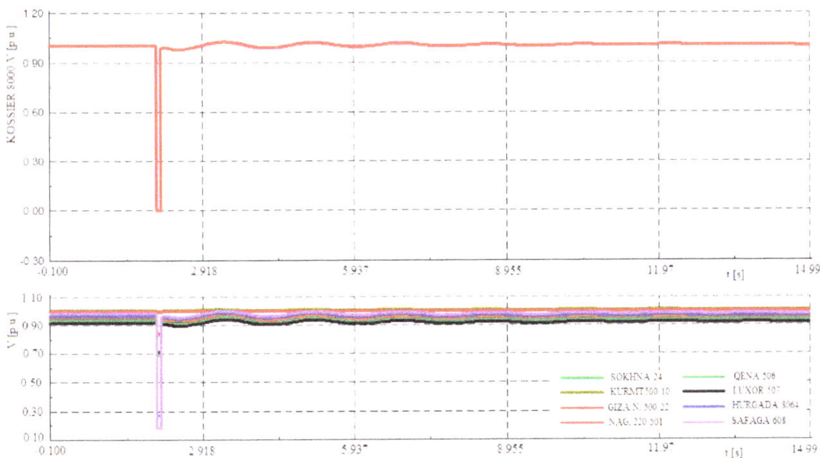

Figure 11. Voltage variation in the case of a 3-phase fault at terminals of PV system at site_2 (KOSSIER) and nearby bus bars; Upper: Terminal voltage at PV terminals, Lower: Terminal voltage at nearby buses.

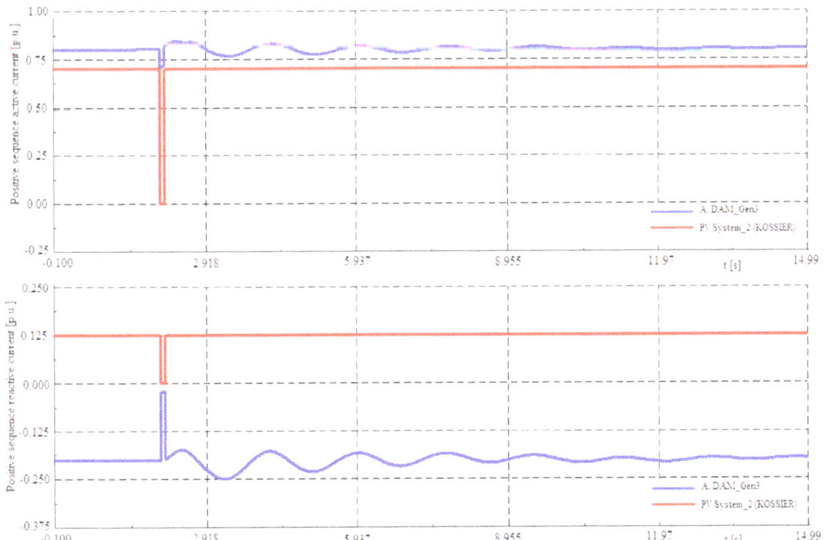

Figure 12. Variations in active and reactive components of current during a 3-phase fault at PV System _2 (KOSSIER) and nearby substations.

3.3.3. Three Phase Short Circuit on the Terminals of PV System_3 (Minia, B. Mazar)

Figure 13 shows the variation in the voltage profile in case of the application of a 3-phase fault for 0.1 s at B. Mazar site in the city of Minia. The upper graph presents the variation in terminal voltage at the terminals of the MINIA power plant. The lower graph shows the terminal voltage variations at surrounding substations shown in the schematic diagram of Figure 4c. Likewise, Figure 14 represents the variations of the active and reactive components of currents injected at B. Mazar site and the thermal power plants of KURM$_1$ and BINI SUIF. The results also indicate that the system is stable. Moreover, it has a good transient performance to rapid recovery of the terminal voltage and reactive and active power after fault clearance.

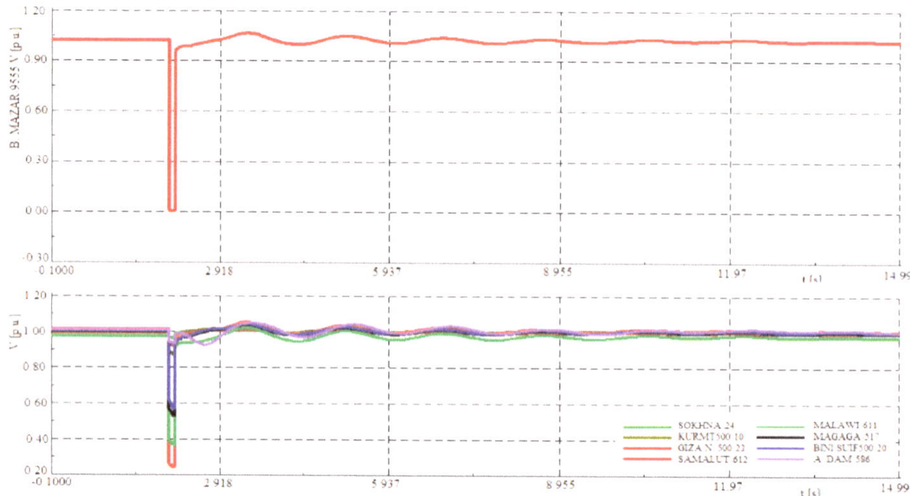

Figure 13. Voltage variation in the case of a 3-phase fault at terminals of PV system at site_3 (B. Mazar) and nearby bus-bars; Upper: Terminal voltage at PV terminals. Lower: Terminal voltage at nearby buses.

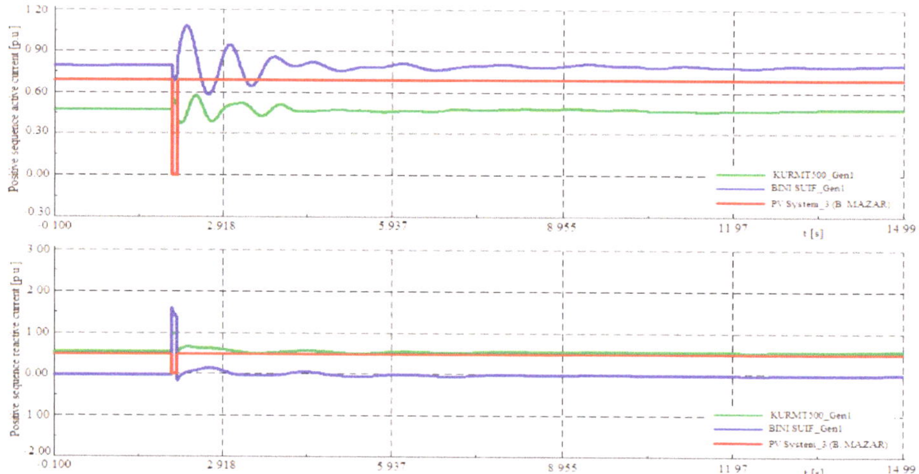

Figure 14. Variations in active and reactive components of current during a 3-phase fault at PV System_3 (B. Mazar, Minia) and nearby substations.

3.3.4. Three Phase Short Circuit on the Terminals of PV System_4 (6.OCT)

Figure 15 shows the variation in the voltage profile in case of the application of a 3-phase fault for 0.1 s at the fourth proposed site for installation of PVPs (6.OCT site). The upper graph presents the variation in terminal voltage at the terminals of the 6.OCT power plant. The lower graph shows the terminal voltage variations at surrounding substations presented in the single line diagram describing this part of the power system given in Figure 4d. Likewise, Figure 16 represents the variations of the active and reactive components of currents injected at 6.OCT site, OCT.GEN. and T.OCT.

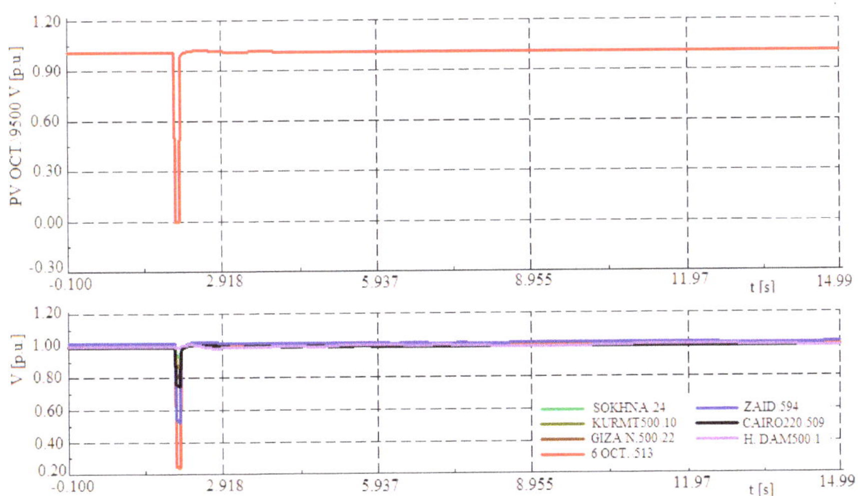

Figure 15. Voltage variation in the case of a 3-phase fault at terminals of PV system at site_4 and nearby bus-bars; Upper: Terminal voltage at PV terminals. Lower: Terminal voltage at nearby buses.

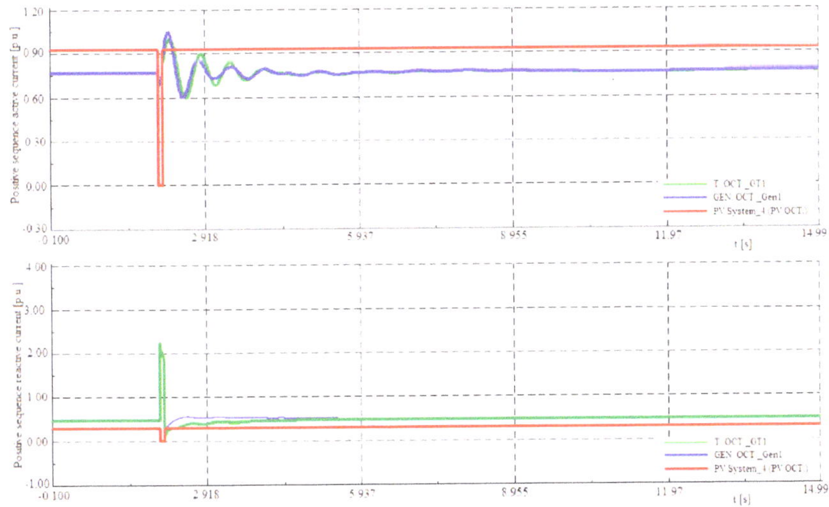

Figure 16. Variations in active and reactive components of current during a 3-phase fault at 6.OCT and nearby substations.

3.4. Frequency Stability Response with the Interconnection of PV Power Plants

In this section, the case of an outage of a certain generating station was simulated to examine frequency stability in the Egyptian power system. The following events have been considered:

(1)　The total demand for energy is covered from conventional power plants with 0% PV penetration,

(2)　3000MW of the load demand is obtained from the proposed PV power plants.

Figures 17 and 18 show the variation of the electrical frequency after an incident tripping of a large generating power plant (NORTH GIZA), with a generating capacity of 2250 MW for the two predefined scenarios.

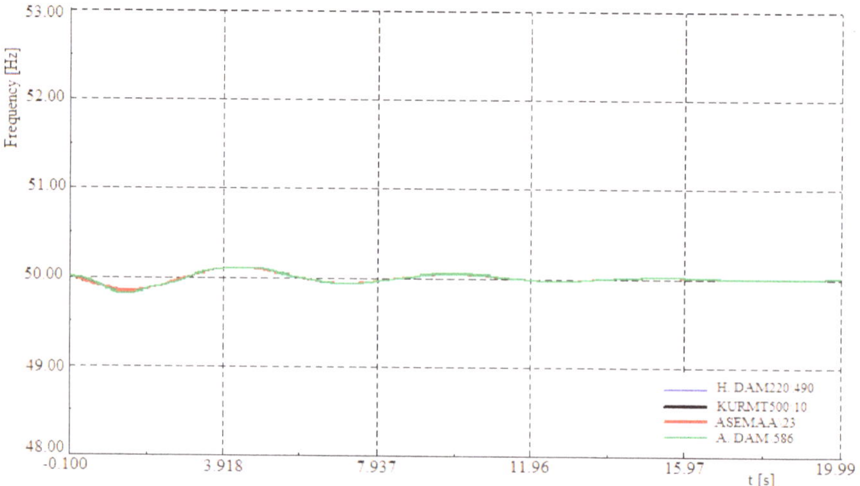

Figure 17. The response of electrical frequency of the power system following the outage of NORTH GIZA plant with 0% PV penetration.

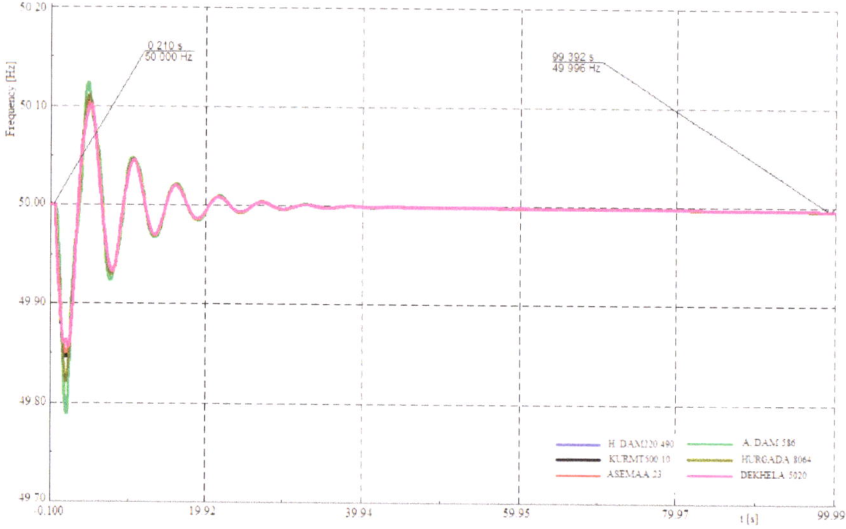

Figure 18. The response of electrical frequency of the power system following the outage of NORTH GIZA plant with 3000 MW from proposed PV sites.

Figures 19 and 20 show the response of rotor angle of a number of the generators of the conventional generating stations operating on the system with No PV generation and with 3000MW renewable energy from the proposed sites of the PV power plants. From these figures, it can be noticed that all generators have reached steady-state condition after the outage of NORTH GIZA power plant.

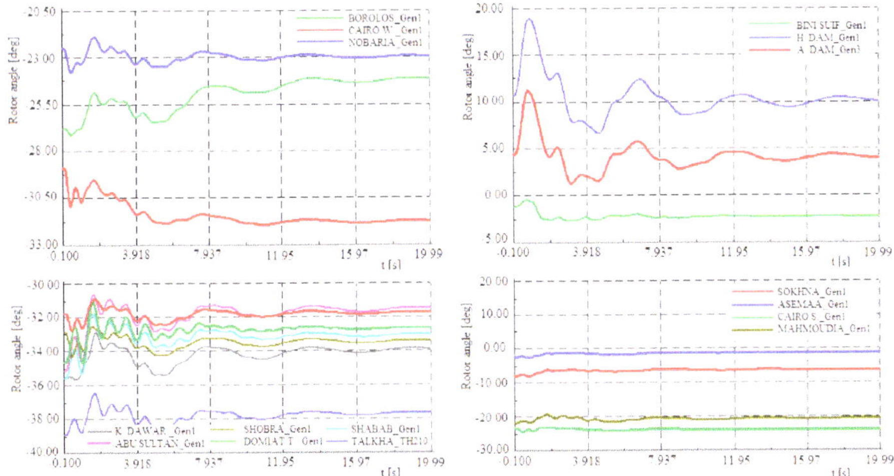

Figure 19. Response of rotor angle of other conventional generating units in the power system with 0% PV penetration.

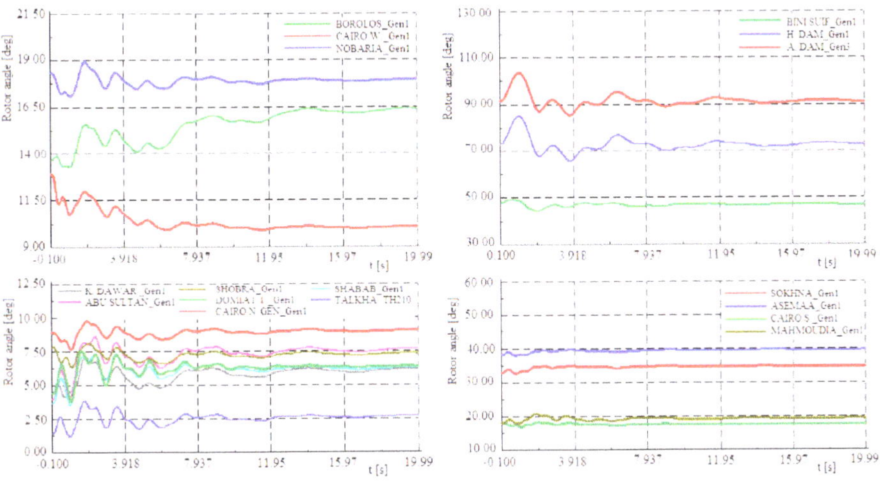

Figure 20. Response of rotor angle of other conventional generating units in the power system with 3000 MW from the PV power plants.

4. Conclusions

The performance of the national utility grid in Egypt is demonstrated using computational simulations using DigSILENT PowerFactory software package in case of high generation levels from photovoltaic power. The impact of high PV penetrations on the capacity of the Egyptian power system is performed using a Newton-Raphson load flow method. In accordance with the constraints on the capacity of transmission lines, the simulation results proved that the maximum allowable generations from the proposed stations have not to exceed 850 MW at KOMOMBO site, 400 MW at KOSSIER site and 1200 MW at B. Mazar, MINIA site to bypass line congestions. Moreover, the capacity of the PVP at 6.OCT site is comparatively high because of the high demand for electric energy at that industrial area and continuous network development. Transmission network amelioration to accept up to 3000 MW from PV plants by 2025 was introduced. Static and dynamic voltage stability of the network has been examined with respect to the integration of large-scale photovoltaic generation.

P-V curves, which have been obtained by increasing the level of PV penetrations at the selected sites, show that voltage control has contributed to raising the level of photovoltaic generation at different nodes of the system through the control of injected reactive power. Furthermore, the study of voltage's dynamic stability, after a three-phase short circuit occurrence, has manifested that the components of reactive power participated in supporting the value of the terminal voltage of the generating units. Analysis of frequency stability has been carried out after the outage of NORTH-GIZA steam power station. It has been observed that frequency stability in the Egyptian utility grid can be maintained for photovoltaic generations up to 3000 MW of the total demand for energy.

Author Contributions: H.M.S. developed the simulation model of the national utility grid of Egypt and analyzed the presented results in this paper under the guidance of A.A.Z.D. and O.N.K. H.M.S., A.A.Z.D. and Z.M.A. performed models simulations, analyzed the data and results and wrote the paper. A.A.Z.D., Z.M.A. and O.N.K. revised and edited the final text of the paper. Z.M.A. and O.A. contributed by drafting and revising. All authors together organized and refined the manuscript in the present form. All authors have approved the final version of the submitted paper.

Funding: This research received no external funding.

Conflicts of Interest: The authors declare no conflict of interest.

References

1. Katiraei, F.; Agüero, J.R. Solar PV Integration Challenges. *IEEE Power Energy Mag.* **2011**, *9*, 62–71. [CrossRef]
2. Qutaishat, S.; Al-Salaymeh, A.; Obeid, H. The dynamic behaviour of large scale Safawi PV plant integrated to the national transmission grid of Jordan, particularly 132 kV busbar. In Proceedings of the Fifth Conference on Renewable and Energy Efficiency for Desert Regions GCREEDER, Aman, Jordan, 4–6 April 2016.
3. Qutaishat, S.; Al-Salaymeh, A.; Obeid, H. Maximum PV penetration level integrated to the national transmission grid of Jordan, particularly 132 kV busbar. In Proceedings of the Fifth Conference on Renewable and Energy Efficiency for Desert Regions GCREEDER, Aman, Jordan, 4–6 April 2016.
4. Patlitzianas, K.D. Solar energy in Egypt: Significant business opportunities. *Renew. Energy* **2011**, *36*, 2305–2311. [CrossRef]
5. Arab Republic of Egypt, Ministry of Electricity and Energy. Egyptian Electricity Holding Company—Annual Report 2015–2016. Available online: http://www.moee.gov.eg/english_new/report.aspx (accessed on 21 November 2018).
6. Grasse, W.; Geyer, M. *Solar Power and Chemical Energy Systems*; International Energy Agency (IEA): Paris, France, 2001; p. 169.
7. The Official Website of the Ministry of Electricity and Energy. Available online: http://www.moee.gov.eg (accessed on 21 November 2018).
8. International Energy Agency. Technology Roadmap Solar Photovoltaic Energy. 2014. Available online: https://www.iea.org/publications/freepublications/publication/TechnologyRoadmapSolarPhotovoltaicEnergy_2014edition.pdf (accessed on 10 February 2019).
9. Tamimi, B.; Canizares, C.; Bhattacharya, K. System stability impact of large-scale and distributed solar photovoltaic generation: The case of Ontario, Canada. *IEEE Trans. Sustain. Energy* **2013**, *4*, 680–688. [CrossRef]
10. Kawabe, K.; Tanaka, K. Impact of dynamic behavior of photovoltaic power generation systems on short-term voltage stability. *IEEE Trans. Power Syst.* **2015**, *30*, 3416–3424. [CrossRef]
11. Moursi, M.; Xiao, W.; Kirtley, J.L. Fault ride through capability for grid interfacing large scale PV power plants. *IET Gener. Transm. Distrib.* **2013**, *7*, 1027–1036. [CrossRef]

12. Daniel, R.; Antoni, C.; Juan, M.; Pedro, R. Power system stability analysis under increasing penetration of photovoltaic power plants with synchronous power controllers. *IET Renew. Power Gener.* **2017**, *11*, 733–741.

13. Kim, S.; Kang, B.; Bae, S.; Park, J. Application of SMES and grid code compliance to wind/photovoltaic generation system. *IEEE Trans. Appl. Supercond.* **2013**, *23*, 5000804.

14. Ruiz, A. System Aspects of Large Scale Implementation of a Photovoltaic Power Plant. MSc Thesis, KTH Electrical Engineering, Stockholm, Sweden, 2011.

15. Tamimi, B.; Cañizares, C.; Bhattacharya, K. Modeling and performance analysis of large solar photovoltaic generation on voltage stability and inter-area oscillations. In Proceedings of the IEEE Power and Energy Society General Meeting, Detroit, MI, USA, 24–29 July 2011.

16. Ghaffarianfar, M.; Hajizadeh, A. Voltage Stability of Low-Voltage Distribution Grid with High Penetration of Photovoltaic Power Units. *Energies* **2018**, *11*, 1960. [CrossRef]

17. Widén, J.; Wäckelgård, E.; Paatero, J.; Lund, P. Impacts of distributed photovoltaics on network voltages: Stochastic simulations of three Swedish low-voltage distribution grids. *Electr. Power Syst. Res.* **2010**, *80*, 1562–1571. [CrossRef]

18. Rahmann, C.; Castillo, A. Fast Frequency Response Capability of Photovoltaic Power Plants: The Necessity of New Grid Requirements and Definitions. *Energies* **2014**, *7*, 6306–6322. [CrossRef]

19. Feilat, E.A.; Azzam, S.; Al-Salaymeh, A. Impact of Large PV and Wind Power Plants on Voltage and Frequency Stability of Jordan's National Grid. *Sustain. Cities Soc.* **2018**, *36*, 257–271. [CrossRef]

20. Ana, C.T.; Oriol, B. Dynamic study of a photovoltaic power plant interconnected with the grid. *ISGTEurope* **2016**, 1–6. [CrossRef]

21. Delille, G.; François, B.; Malarange, G. Dynamic frequency control support by energy storage to reduce the impact of wind and solar generation on isolated power system's inertia. *IEEE Trans. Sustain. Energy* **2012**, *3*, 931–939. [CrossRef]

22. Eftekharnejad, S.; Vittal, V.; Heydt, T.G.; Keel, B.; Loehr, J. Impact of increased penetration of photovoltaic generation on power systems. *IEEE Trans. Power Syst.* **2013**, *28*, 893–901. [CrossRef]

23. Khan, M.; Arifin, M.; Haque, A.; Al-Masood, N. Stability analysis of power system with the penetration of photovoltaic based generation. *Int. J. Energy Power Eng.* **2013**, *2*, 84–89. [CrossRef]

24. Fetouh, T.; Kawady, T.A.; Shaaban, H.; Elsherif, A. Impact of the new Gabl El-Zite wind farm addition on the Egyptian power system stability. In Proceedings of the 13th International Conference on Environment and Electrical Engineering (EEEIC), Wroclaw, Poland, 1–3 November 2013; pp. 96–103. [CrossRef]

25. Gado, A.E. Impact the expansion of the production of generation of solar power on the low voltage network in Egypt. In Proceedings of the 2015 Saudi Arabia Smart Grid (SASG), Jeddah, Saudi Arabia, 7–9 December 2015; pp. 1–4.

26. Tageldin, E.M.; Elghazaly, H.M.; Moussa, M.M. An Integrated Protective Scheme for a multi-ended Egyptian Transmission Line using Radial Basis Neural Network. In Proceedings of the 6th International Conference on Power Systems Transients, Montreal, QC, Canada, 20–23 June 2005.

27. Gonzalez-Longatt, F.; Torres, J.L.R. *Advanced Smart Grid Functionalities Based on Power Factory*; Springer International Publishing: Cham, Switzerland, 2018.

28. Gonzalez-Longatt, F.; Torres, J.L.R. *Power Factory Applications for Power System Analysis*; Springer: Cham, Switzerland, 2014.

29. Ab Wahab, L.H. Modeling Study Power Distribution System Network Using DIgSILENT PowerFactory 13.2 Simulation Software. Bachelor's Thesis, University Malaysia Pahang, Pekan, Malaysia, November 2009.

30. Zhu, Y.; Tomsovic, K. Development of three-phase unbalanced power flow using PV and PQ models for distributed generation and study of the impact of DG models. *IEEE Trans. Power Syst.* **2007**, *22*, 1019–1025.

31. Wirasanti, P.; Egon, O. Active Distribution Grid Power Flow Analysis using Asymmetrical Hybrid Technique. *Int. J. Electr. Comput. Engin. (IJECE)* **2017**, *7*, 1738–1748. [CrossRef]

32. Shotorbani, A.M.; Madadi, S.; Mohammadi-Ivatloo, B. Wide-Area Measurement, Monitoring and Control: PMU-Based Distributed Wide-Area Damping Control Design Based on Heuristic Optimisation Using DIgSILENT PowerFactory. Advanced Smart Grid Functionalities Based on PowerFactory. In *Advanced Smart Grid Functionalities Based on PowerFactory. Green Energy and Technology*; Gonzalez-Longatt, F., Rueda Torres, J., Eds.; Springer: Cham, Switzerland, 2018; pp. 211–240.

33. Hammad, M.; Ahmad, H. Static analysis for voltage stability of the Northern Jordanian power system. In Proceedings of the 9th International Renewable Energy Congress (IREC), Hammamet, Tunisia, 20–22 March 2018.

34. Wang, C.; Yuan, K.; Li, P.; Jiao, B.; Song, G. A projective integration method for transient stability assessment of power systems with a high penetration of distributed generation. *IEEE Trans. Smart Grid* **2018**, *9*, 386–395. [CrossRef]

35. Schinke, A.; Erlich, I. Enhanced Voltage and Frequency Stability for Power Systems with High Penetration of Distributed Photovoltaic Generation. *IFAC-Papers OnLine* **2018**, *51*, 31–36. [CrossRef]

36. Shah, R.; Robin, P.; Mike, B. The Impact of Voltage Regulation of Multi-infeed VSC-HVDC on Power System Stability. *IEEE Trans. Energy Conv.* **2018**, *33*, 1614–1627. [CrossRef]

37. Kusumo, S.A.; Lesnanto, M.P. Transient Stability Study in Grid Integrated Wind Farm. In Proceedings of the 5th International Conference on Information Technology, Computer, and Electrical Engineering (ICITACEE), Semarang, Indonesia, 26–28 September 2018.

38. DIgSILENT. *PowerFactory Users Manual*; DIgSILENT: Gomaringen, Germany, 2017.

39. Hamdy, M.S.; Kuznetsov, O.N.; Ahmed, A.Z.D. Modelling and performance evaluation of the Egyptian national utility grid based on real data. In Proceedings of the IEEE Conference of Russian Young Researchers in Electrical and Electronic Engineering (EIConRus), Moscow, Russia, 29 January–1 February 2018; pp. 807–812.

40. Transmission Grid Code, Egyptian Electricity Transmission Company, EETC, Cairo, Egypt. Available online: http://www.eetc.net.eg/grid_code.html (accessed on 10 February 2019).

41. *1547.2-2008–IEEE Application Guide for IEEE Std 1547(TM), IEEE Standard for Interconnecting Distributed Resources with Electric Power Systems*; IEEE: Piscataway, NJ, USA, 2003; pp. 1–16.

42. Hamdy, M.S.; Kuznetsov, O.N.; Ahmed, A.Z.D. Site Selection of large-scale Grid-connected solar PV system in Egypt. In Proceedings of the IEEE Conference of Russian Young Researchers in Electrical and Electronic Engineering (EIConRus), Moscow, Russia, 29 January–1 February 2018.

43. *IEEE Power Engineering Society Administrative Committee of the Power System Dynamic Performance Committee. Causes of the 2003 Major Grid Blackouts in North America and Europe, and Recommended Means to Improve System Dynamic Performance*; IEEE: Piscataway, NJ, USA, 2005; pp. 1922–1928.

44. International Electrotechnical Commission. *Introduction to IEC 909 (BS 7639). Short-Circuit Current Calculation in Three-Phase AC Systems*, 1st ed.; International Electrotechnical Commission: London, UK, 1988.

45. Taylor, C.W. *Power System Voltage Stability*; McGrawhil: New York, NY, USA, 1993.

46. Salma, A.; Eyad, F.; Ahmed, A.-S. Impact of Connecting Renewable Energy Plants on the Capacity and Voltage Stability of the National Grid of Jordan. In Proceedings of the 8th International Renewable Energy Congress (IREC 2017), Amman, Jordan, 21–23 March 2017. [CrossRef]

Review

Concentrating Solar Power Technologies

Maria Simona Răboacă [1,2,*], **Gheorghe Badea** [2], **Adrian Enache** [1], **Constantin Filote** [3],
Gabriel Răsoi [1], **Mihai Rata** [3], **Alexandru Lavric** [3] **and Raluca-Andreea Felseghi** [2,3,*]

[1]   National Research and Development Institute for Cryogenic and Isotopic Technologies—ICSI Rm. Valcea,
      Uzinei Street, No. 4, P.O. Box 7 Raureni, 240050 Rm. Valcea, Romania; adrian.enache@icsi.ro (A.E.);
      gabi.rasoi@icsi.ro (G.R.)
[2]   Technical University of Cluj-Napoca, Memorandumului Street, No. 28, 400114 Cluj-Napoca, Romania;
      gheorghe.badea@insta.utcluj.ro
[3]   Faculty of Electrical Engineering and Computer Science, Stefan cel Mare University of Suceava,
      720229 Suceava, Romania; filote@usm.ro (C.F.); mihair@eed.usv.ro (M.R.); lavric@eed.usv.ro (A.L.)
*   Correspondence: simona.raboaca@icsi.ro (M.S.R.); Raluca.FELSEGHI@insta.utcluj.ro (R.-A.F.)

Received: 3 February 2019; Accepted: 12 March 2019; Published: 18 March 2019

Abstract: Nowadays, the evolution of solar energy use has turned into a profound issue because of the implications of many points of view, such as technical, social, economic and environmental that impose major constraints for policy-makers in optimizing solar energy alternatives. The topographical constraints regarding the availability of inexhaustible solar energy is driving field development and highlights the need for increasingly more complex solar power systems. The solar energy is an inexhaustible source of CO_2 emission-free energy at a global level. Solar thermal technologies may produce electric power when they are associated with thermal energy storage, and this may be used as a disposable source of limitless energy. Furthermore, it can also be used in industrial processes. Using these high-tech systems in a large area of practice emboldens progress at the performance level. This work compiles the latest literature in order to provide a timely review of the evolution and worldwide implementation of Concentrated Solar Power—CSP—mechanization. The objective of this analysis is to provide thematic documentation as a basis for approaching the concept of a polygeneration solar system and the implementation possibilities. It also aims to highlight the role of the CSP in the current and future world energy system.

Keywords: concentrated solar power (CSP), installed capacity; solar energy resources; solar thermal plants; thermal energy storage (TES)

---

1. Introduction

Prefacing the improvement of inexhaustible energy supply worldwide and non-polluting power sources [1,2], represents one of the major purposes of power generation at a global level [3–5]. Renewable energies (sunlight, sunscreen, water strength, biomass, wind power) and renewable raw materials are alternatives to fossil resources [3]. Solar activity represents one of the purest types of energy [6]. The huge amount of this type of energy underlies almost all natural processes on Earth [7]. It is, however, quite difficult to capture and store in a usable form (mainly heat or electricity) [8,9] that would facilitate its subsequent use [10].

The potential of solar resources, which far exceeds the potential of fossil fuels, is given by the following characteristics [1]:

-   Solar resources are inexhaustible (the Sun provides the Earth with 15,000 times more energy than the annual energy consumption of atomic or fossil energy; the solar source can supply the Earth's energy needs for at least five billion years);

- Solar resources are wholly or partly available everywhere, assuming regional decentralized operation;
- The transformation of solar resources into secondary energy and in secondary materials like heat, fuel, electricity does not emit CO_2 not shielding the global environment;
- There is the possibility of developing a sustainable civilization model [1].

Solar Thermal Energy (STE) is the most important type of solar power activity and represents one of the important technology resources, producing energy useful in various applications such as: building, electromobility and manufacturing. The integration with thermal energy storage (TES) comes with the possibility to make STE unique and dispatchable when mixed with other inexhaustible sources of energy. For a long period, expansion of the thermal solar activity industry has been associated with TES theory. It is important to provide high tech sources which facilitate the distribution of the demanded energy supply [11].

Unlike photovoltaic (PV) panel technologies, Concentrated Solar Power (CSP) has an inherent capacity to store heat energy for limited intervals of time for later conversion into electricity. When combined with thermal storage capacity, CSP plants are able to produce electricity even when clouds block the Sun or after sunset. [12]. Additionally, for instance, one megawatt of installed CSP avoids the emission of 688 tons of CO_2 compared to a combined cycle system, and 1360 tons of CO_2 compared to a coal/steam cycle power plant. One square mirror in the solar field produces 400 kWh of electricity per year, avoids 12 tons of CO_2 emission, and contributes to 2.5 tons savings of fossil fuels during 25-year operation life time [13].

This work aimed to provide a state-of-the-art review of the development of CSP technologies over the last decade. First, the article provided a summary on the status of the EU's main objectives for renewable energy sources (RES) development, which intended to highlight the role of the CSP in the current and future energy system of Europe. The main CSP technologies are presented and the suitability map for installation of solar thermal power plants according to the direct normal irradiance (DNI) was illustrated. There is discussion regarding the worldwide stage of installation of capacities based on CSP technology, as well as CSP plants that are currently operating or under construction. The goal of this review was to provide a thematic documentation that can be a starting point for developing a research project within National Research and Development Institute for Cryogenic and Isotopic Technologies—ICSI Rm. Valcea, ICSI Energy Department. This work has already achieved results in some "smart city" examples in Europe [14]. It aimed to approach the concept of a polygeneration solar system, which involved the possibility of obtaining several forms of useful energy from solar resources: electricity, thermal energy (heat), mechanical work produced by steam, chemical energy in the form of hydrogen (fuel), cooling energy, light flux, etc.

Romania, one of the EU members, through energy policies, adopted strategies and research activities [15], has an intense orientation towards the world solar economy [16], which requires a revolution in energy technology, making the technical development of productive forces replicable internationally [1].

2. Materials and Methods

To compile the review based on a literature research of Concentrated Solar Power (CSP) technologies for sustainable power generation, existing relevant studies that were analyzed based on different types of CSP along with thermal energy storage (TES) technologies, and the worldwide state of implementation of these concepts have been identified.

A systematic literature search was carried out in Science Direct, MDPI, ResearchGate, Google Scholar and specialized technical platforms to identify relevant studies involving review analysis of different types of CSP and TES technologies and installed capacities, during the last 10 years. The concept of CSP technologies is not new, and a significant number of studies have already been conducted by researchers. The identified research works have been characterized according to the technologies reviewed, methodology adopted and the sustainability parameters discussed.

3. Considerations Regarding the CSP Technologies

3.1. Main Policies and Objectives for Renewable Energy Sources (RES) Development

The power of the sun and of the wind, as well as the power provided by biomass and biofuels, along with geothermal and hydro power energy are used as alternatives for fossil fuels to avoid the emissions that can trigger the greenhouse effect, and reducing the dependency volatility for fossil fuels, especially oil and gas [4,17,18].

The future post 2020 timeframe study, as well as the EU legislation on the promotion of renewable sources is under debate. The fundamental process of society's development is based on the availability of an inexhaustible power supply [3,19]. Contributing to most of the proposed energy requires, ensuring the transition to a sustainable energy system, the security of supply, reduction and even elimination of the greenhouse gas emissions, and the industrial development, that would lead to job growth and significantly lower energy costs [20]. The objectives of "20-20-20" strategy to be fulfilled by 2020 have set the following three key targets:

- 20% reduction in EU greenhouse gas emissions compared to 1999 levels;
- 20% increase in the share of energy produced from renewable sources in the EU;
- improving the energy efficiency in the EU by 20% [15].

As the European energy system faces an increasingly pressing need for sustainable, affordable and competitive energy for all citizens, the European Commission adopted on 30 November 2016 the legislative package "Clean Energy for All Europeans", which seeks to implement strategies and measures to achieve the objectives of the Energy Union for the first ten-year period (2021–2030), in particular for the EU's 2030 energy and climate objectives, and refers to: energy security, energy market, energy, de-carbonization, research, innovation and competitiveness [15,19,20].

In a wider perspective, the EU established a set of long-term objectives in roadmaps by 2050. Regarding the building sector, the main three roadmaps are:

- The EU's objective of moving to a low-carbon, competitive economy by 2050, which identified the need to reduce carbon emissions in the residential and service sectors (generically referred to as the real estate sector) by 2050 compared to 1990 levels;
- Energy Perspective 2050, where "increasing the energy efficiency potential of new and existing buildings is essential" for a sustainable future;
- The Energy Efficient Europe Plan, identifying the real estate sector as one of the top three sectors responsible for 70% to 80% of the total negative environmental impact. Achieving better construction and optimizing their use within the EU would reduce by over 50% the amount of raw materials extracted from the underground and could reduce water consumption by 30%.

These roadmaps are a long-term aspiration that is not only desirable from a social and economic point of view, but also ecologically essential [15,19,20]. In many countries, the strong development of the heating sector from renewable sources [21] has been a key factor in achieving and surpassing the intermediate targets in these EU member states. This is true, for example, in Bulgaria, Finland and Sweden, where development was mainly driven by the use of low-cost fuel from biomass. The use of an inexhaustible supply in the field of transport [22] has lagged behind in most countries, except for Sweden, Finland, Austria, France and Germany.

However, most Member States are about to meet and even surpass their targets by 2020, based on the planning and assessment of current policies [3,19,20]. This makes it an entirely possible target not only for the EU members, but also for the entire EU (Figure 1) [15].

Solid biomass fuels were the main factor that produced heat from inexhaustible sources in 2013 [23], while during 2014, a report was published in regards to the production of heat and electricity based on the solid sustainable and gaseous biomass. This report was published by the European Union and contained information on current planned EU actions that were supposed to maximize the benefits of the biomass usage while avoiding negative impacts on the environment [15].

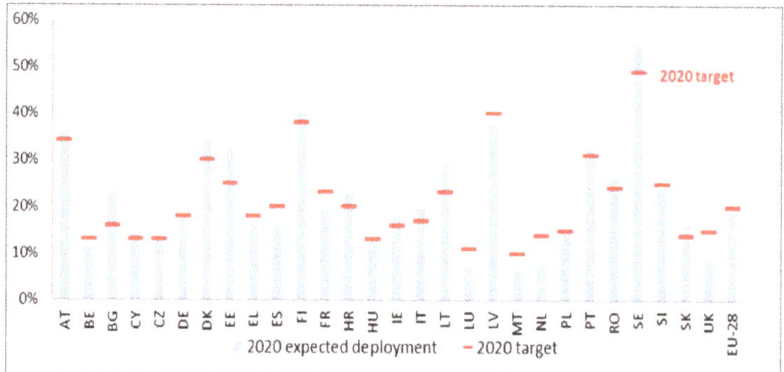

Figure 1. Implementation chart of renewable energy sources (RES) in the EU member states and 2020 targets [15].

The power plants based on concentrated solar radiation (CSP) are considered to be an interesting alternative for generating electricity from renewable energy on a large scale worldwide [24,25]. Although their development has not been so rapid, some relevant projects were still in pending or cancelled in countries like Australia, and expansion of solar heat supply capacity is expected to be taken into consideration in the following years [26].

Electricity power, the fastest growing form of energy [27], is the power sector that contributes more than any other to the reduction of fossil fuels worldwide [28]. By 2040, an increase of around 40% (based on the current fleet) of the electricity requirement is expected. This increase by 7200 gigawatts may substitute the existing energy generators [5,29]. The strongest growth of renewables in many countries raises their share worldwide power generation, to one third by 2040 [15,30].

Due to global population industrialization and growth, the energy demand has increased dramatically. We can obtain renewable energy through natural resources such as geothermal, biomass, wind and solar heat power. This is the focus of most countries, especially due to the sustainability benefits by reduction of CO_2 and greenhouse gas emissions.

Among the technologies based on the use of solar sources to produce energy, the technology using parabolic mirrors outstanding by high efficiency, compactness and the advantage of modularisation that allows them to be placed in isolated places, independent of access to conventional energy sources. This technology is part of the radiation focusing category, the mirrors having a role of capturing and concentrating incident solar rays in the focal area where the receiver is located, which is usually coupled to an electric generator. Thus, solar energy is transformed into thermal energy into the receiver, then to mechanical energy in the engine and finally to electricity in the generator [31]. Coupling of technologies through integration of thermal energy storage (TES) to concentrating solar power (CSP) brings uniqueness to this type of power converter among all other renewable energy generating alternatives [30,31].

3.2. Direct Normal Irradiance (DNI)

Direct Normal Irradiance (DNI) is defined as the solar irradiance collected by a normal plane, directly from the Sun, being of high importance, respectively is the basis of the functioning principle of CSP technology [32–34].

Figure 2 presents a map of the global distribution of direct normal irradiation, where four zones can be further distinguished [35–37] based on their suitability for the installation of solar thermal power plants (Figure 3) [37]. Thus, zone I—excellent—has great potential and records maximum DNI values of 3652 kWh/m^2/y. It is followed by zone II—having a good potential for exploitation of DNI, which has average values of 2800 kWh/m^2/y. Zone III allows to install thermodynamic solar power plants with DNI values of 1700 ÷ 2100 kWh/m^2/y. And, finally, area IV—not suitable for these types of energy generation systems that have a DNI of 365 ÷ 1700 kWh/m^2/y [37,38].

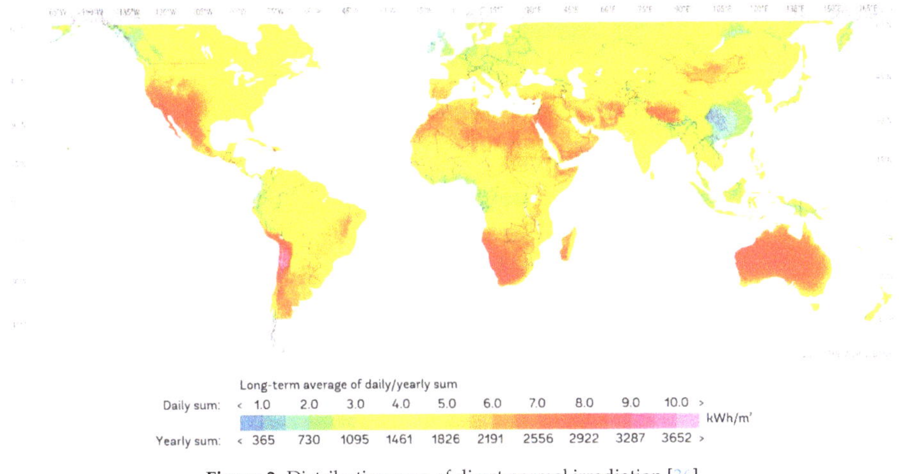

Figure 2. Distribution map of direct normal irradiation [36].

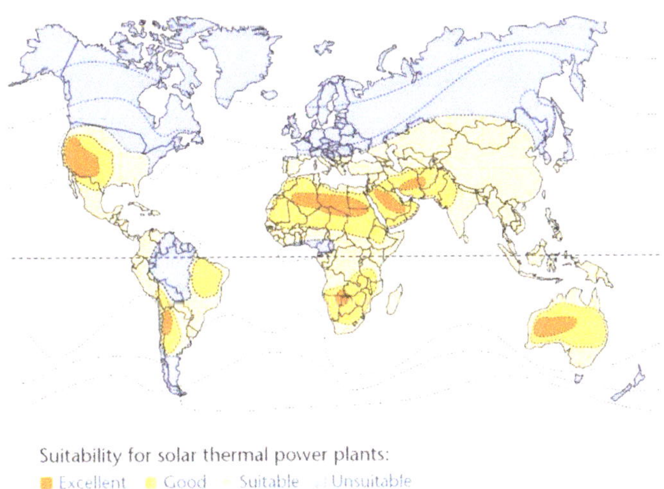

Figure 3. Suitability map for installation of solar thermal power plants [37].

For an efficient operation of thermal solar plants, DNIs need to record values above 1800 kWh/m^2/y [39]. The most favorable CSP resource [37] areas are thus in North Africa, the south-western United States, northern Mexico, north-western India middle East, southern Africa, Chile, Australia, Peru and the western parts of China. Other relevant areas such as southern areas from Europe, Turkey and US, central area of Asia, Brazil and also Argentina and China, are included [40,41].

3.3. The Concept of Concentrated Solar Power (CSP)

One of the first studies on the possible use of sunlight, dating back to 1774, in the late 18th century, belongs to Antoine Lavoisier who created a large optical device containing a glass lens to focus and concentrate the sunlight on the surface of a burning material. Later, in 1878, a parabolic collector was designed and built to test the impact of the Sun's rays on a steam boiler to heat the water from its interior to the boiling point, and to release steam under pressure. This boiler by means of a mechanical device, was ran and powered a printing press. [42,43] Figure 4a presents the first concept of a solar parabolic collector.

Sophisticated solar power (CSP) technologies are currently under development but are still not as accessible as conventional photovoltaic panels in providing confidence and reliability [44,45]. Figure 4b presents a modern solar parabolic concentrator concept [46].

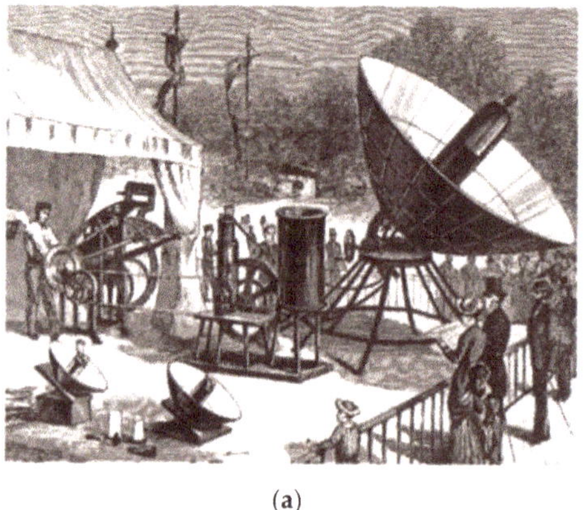 

(a)   (b)

Figure 4. Solar parabolic collector [43,46]. (a) First concept; (b) Modern concept.

Solar energy concentrating systems use parabolic mirrors that reflect the sunlight on a single point over the receiver's surface from where it is collected and transformed into thermal and electrical energy. Parabolic mirrors are designed to focus solar radiation on the receiver, which heats the gas to a relatively high temperature, which is then used to move a turbine or steam to power a Stirling engine running an electric generator, thus producing electricity [31,43].

Thermodynamic solar systems put into operation optical concentrators that exploited direct sunlight. The main specific of CSP technology(illustrated in Figure 5), when compared to other renewable energy conversion equipment is represented by a thermic stock system that generates electrical power during intervals of time with cloudy skies or the sun setting.

As compared to photovoltaic panels (PV), CSP uses DNI in order to provide heat and electricity without CO_2 emissions where the DNI level is higher in comparison with others [43,47,48]. The CSP commercial technologies are the following [39,49–52]:

(a) *Linear concentrating systems* which include parabolic troughs and linear Fresnel reflectors: Parabolic Parabolic Troughs (PTs) usually count on oil as synthetic fuel to facilitate an exchange of power from the collector pipes into heat. During this process, the water is boiled and it evaporates, running the turbine and driving the plant to create electric energy. On a commercial basis, we can say that CSP is proved to be the most established technology. Linear Fresnel reflectors (LFR) make up a series of ground-based flat mirrors placed at angles that help concentrate the sunlight, in order to locate a receiver from several meters above. Compared to PT, the LFR shows a lower performance [39,49–52].

(b) *Solar Power Towers (STs)* increase the number of computer-assisted mirrors to track the Sun in an individual way over two axes. In this way it concentrates the solar irradiation onto a single angle which we can fix on the top on the tower, placed in a central point; there, the heat produced by the sun conducts a thermodynamic cycle and produces electricity, so that the ST plants can approach higher temperatures than the other two above mentioned systems (PT and LFR) [39,49–52].

(c)   *A Parabolic Dish (PD)* is made up of a parabolic dish-shaped concentrator that mirrors the Direct Normal Irradiance into a receiver located at the focal point of the dish. The main advantages of PD technologies include high energy efficiency (up to 30%) and modularity (5–50 kW), in addition to being particularly suited to distributed generation systems [39,49–52].

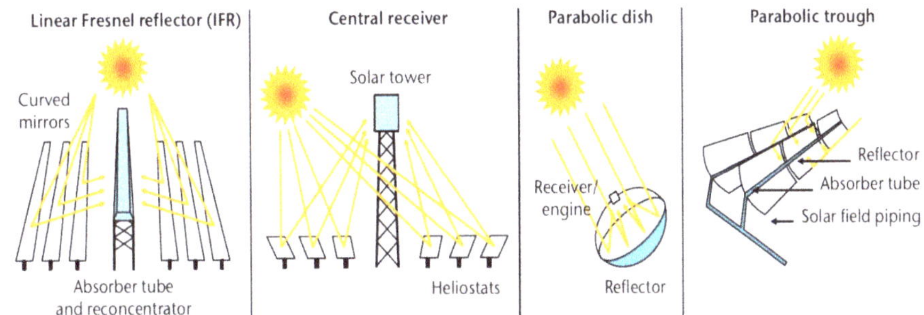

Figure 5. Main CSP technologies [53].

Based on the literature [54–62], the specific features of CSP technologies are presented in a comparative manner in Table 1.

Table 1. Comparison of CSP technologies [54–62].

CSP Type	Operating Temperature (°C)	Ratio of Solar Concentration	Thermal Storage Suitability	Average Annual Efficiency	Land Use Efficiency (Total Area/Power)
Parabolic Trough	20–400	15–45	Suitable	15%	3.9
Linear Fresnel Reflector	50–300	10–40	Suitable	8–11%	0.8–1
Solar Trough	300–1000	150–1500	Highly suitable	17–35%	5.4
Parabolic Dish	120–1500	100–1000	Difficult	25–30%	1.2–1.6

PT systems occupy a large area, having low thermodynamic efficiency due to their low operating temperature. They have a relatively low installation cost and a large experimental feedback. Furthermore, LFR has thermodynamic efficiency due to the low operating temperature, but low installation cost. ST has high thermodynamic efficiency due to a high operating temperature; it occupies a large area, it has high installation cost and records high heat losses. Finally, PD occupies small area, features high thermodynamic efficiency due to high operating temperature, but it requires a high installation cost [54–62].

A percentage of 85% of all CSP research projects is carried out on the parabolic troughs technology; accordingly, the existing data regarding operating experience and cost information generally refers to PT energy systems.

Levelized Cost of Electricity (LCOE) estimated for CSP is still high as compared to the other renewable technologies, as shown in Table 2 [55,63–66].

Power generating technologies based on alternative energies are subject to constant research and development. Costs are expected to decline in the near future due to various pilot projects currently underway in this field being validated and implemented on a large scale (including energy storage solutions).

Table 2. LCOE estimates (€c/kWh) [55,63–68].

Technology	Europe	USA	China
CSP	17.6–43.10	17.6–43.10	16.70–40.50
PV	8.80–22.00	9.70–20.25	6.95–13.15
Wind	6.25–10.30	5.40–11.95	4.30–8.20
Hydro	8.80	7.95	2.65
Coal	10.55–14.95	6.20	3.10–3.45

4. Discussion Regarding the Stage Installation of Capacities based on CSP

4.1. General Considerations on RES

Figure 6 shows a country ranking on total renewable energy production capacity as of the end of 2017 [68].

Figure 6. Country ranking on total renewable energy production capacity for 2017 [68].

From Figure 6, it can be seen, that the countries with the largest renewable energy production are China, the US and Brazil, followed by Germany and India. In terms of power generated by CSP capacity, Spain ranks first in the world ranking, being followed by United States that ranks second, and then South Africa, India, Morocco and China. In terms of solar water heating collector capacity, China ranks first in the world ranking, being followed by United States, Turkey, Germany and Brazil.

4.2. Worldwide Capacity of CSP Technologies

Figure 7 shows the concentrating solar thermal power global capacity by country and region over the period 2007–2017. From the statistical data presented in Figure 7, there can be observed a significant increase in renewable energy capacity from 0.4 gigawatts produced in 2007 to 4.9 gigawatts in 2017. Despite the fact that the global capacity increased by only over 2% in 2017 compared to 2016, the CSP industry was active, with a pipeline of about 2 GW of projects under construction around the world, especially in the Middle East and North Africa (MENA) region and in China [37].

Concentrated solar power technologies (CSP) have helped boost developing countries with a high level of direct normal irradiation (DNI) [35] and a specific strategic and/or economic alignment, benefitting from the advantages of these technologies [40]. CSP technologies benefit from better support for energy policies, low oil and gas reserves with limited access to electricity grids, or stringent energy storage needs, thus achieving a strong industrialization and creating new jobs [33].

Ongoing research conducted mainly in Australia, Europe, and the United States, has kept concentrating on the development and improvement of Energy Storage Technology (TES) [69].

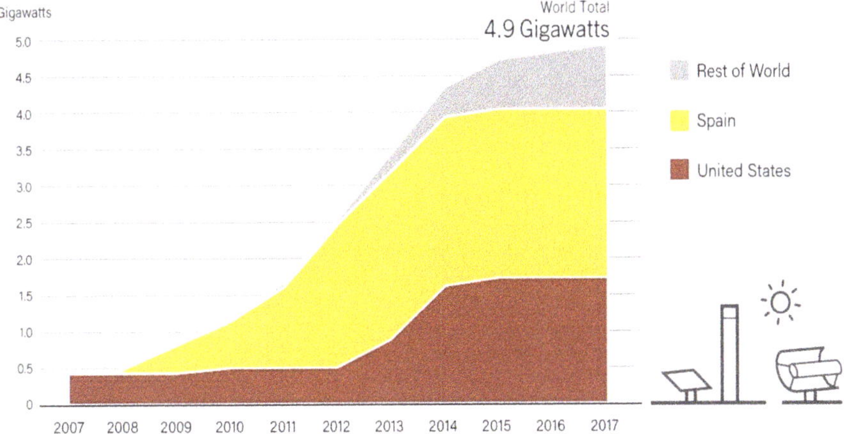

Figure 7. Concentrating solar thermal power global capacity by country and region over the period 2007–2017 [68].

Figure 8 shows the global storage capacity of solar thermal energy during the period 2007–2017. According to the statistical data presented in Figure 8, there is significant increase in the global storage capacity of thermal energy produced from concentrated solar radiation, from the supply of about 0.04 megawatts-hours in 2007 to 12.8 gigawatts-hours, in 2017.

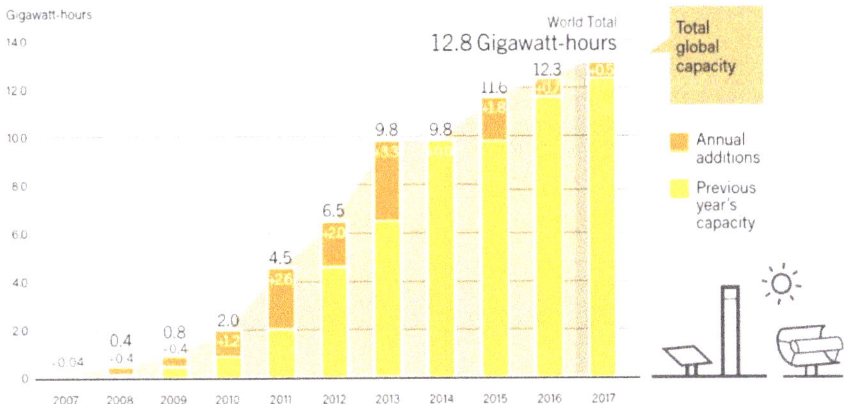

Figure 8. CSP thermal energy storage global capacity and annual additions, 2007–2017 [70].

Figure 9 shows the STE worldwide capacity organized by main CSP technologies [71]. A significant percentage of installations in operation or under construction have linear gradient concentrating systems such as parabolic troughs, which operate both with and without storage. The trend was to increase the use of solar-tower technology. Forty one (41) percent of thermal energy storage systems and 41% of all STE plants under construction are in development under Fresnel reflectors, which operate without storage.

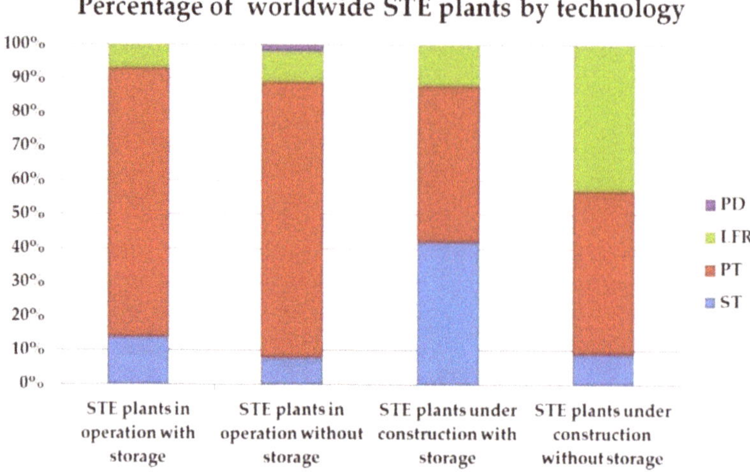

Figure 9. STE worldwide capacity categorized by technology and with/without storage [72,73].

Parabolic dishes are present in a low percentage of approximately 2% of the total STE plants in operation without storage, this type of technology being in the stage of research and implementation at the level of practical applications [72,73].

5. CSP Sunflower 35 Experimental Equipment

Under the research project "ROM-EST: Research Laboratories for Energy Storage", following the research study, a Sunflower 35 solar concentrator was purchased together with a Stirling energy unit to produce energy and heat from concentrated solar energy.

The performance of the Sunflower 35 module according to the manufacturer's specifications is as follows:

- Stirling engine type α, maximum power of 10 kWe at 1500 rpm, useful volume 183 cm^3, maximum yield 25%.
- Maximum power 10 kWe + 25 kWt (for average direct radiation of 1000 W/m^2), ~400 V AC/ 50 Hz/3 phases, IP55, min.
- Max. 600–700 °C.
- Average working pressure is 15 MPa, helium working gas, engine volume 2 L.
- PCU weight: 500 kg.
- Thermal agent control system with 50 °C inlet cooler water temperature and 60 °C hot water outlet temperature from the cooler.
- electrical power in operation: 2–10 kWel;
- maximum thermal flow (in cogeneration) to direct normal radiation> 1000 W/m^2: about 25 kWt;
- thermal flow in operation (in cogeneration system): 9–25 kWterm;
- simultaneous supply of heat and electricity up to 35 kW;
- total annual output of up to 85 MW;
- high efficiency: 25% for power generation, 70% for cogeneration;
- up to 3600 operating hours per year;
- about 26 MW of electric power supplied annually;
- about 59 MW of heat supplied annually;
- completely dispatchable generation;
- control of the level of power generated;

-       precision of the Sun tracking system.

The Stirling engine did not require water consumption for power generation or for cooling cycles (closed circuit operation). The Stirling Technology Conversion Unit (PCU 35) of 35 kW was equipped with a 25% efficiency SBT V-183 thermal engine.

This equipment is sturdy and durable, adapts to and resists extreme operating conditions such as temperatures from $-20\,^{\circ}\mathrm{C}$ to $+50\,^{\circ}\mathrm{C}$, 100% humidity, snow, sand, wind and dust.

The installation of the solar concentrator together with the power unit and the necessary accessories were initially underwent a preliminary study, which consisted in positioning simulation procedure, which was a preconception of the actual placement and installation mode in the location established. Taking into account these aspects, the installation and assembly positions of the main equipment will be anticipated in a preliminary manner.

The surface requirement for the solar concentrator is about 10×20 m^2, with a favorable position, so as to ensure maximum solar radiation at the lowest shading factor. During the first stage of the installation and commissioning phase, the solar concentrator was installed together with the technical annex for the control and control of the processes.

At a later stage, the power unit with all necessary accessories will be mounted on the support arm of the concentrator, after prior testing of the equipment.

For the safety and proper functioning of the equipment, they must be mounted and installed in a safe location, where they must operate at a constant temperature and protected from bad weather.

For this purpose, a technical annex for the installation of technical equipment for process control and control is placed near the installation, being protected against dust and impurities, with a controlled thermal regime throughout the seasons.

The annex will be divided into two rooms with the following use:

-       installation and mounting of the equipment for thermal energy storage;
-       installing and mounting of the electricity storing equipment.

Figure 10 shows the installation of the solar concentrator working position together with the technical annex for command and control.

Figure 10. Installed solar concentrator in its working position.

The parabolic metal frame of the solar concentrator was designed to be assembled to the ground by means of a special mounting system that allows quick and easy installation of the building.

The support pillar was positioned vertically in the reinforced concrete foundation on which the solar concentrate drive system was mounted, which is shown in Figure 11.

Figure 11. Solar concentrator support pylon.

After mounting the support pillar, the parabolic metal frame assembly was positioned by the arm and secured to the propulsion system.

Install the control system with all the quick connect connections that are installed on the control board interface of the propulsion system. The Stirling motor assembly of the test equipment is presented in Figure 12.

Figure 12. The Stirling motor assembly.

6. Conclusions

There is no doubt regarding the great global potential of solar energy which is a clean renewable energy form. It has the disadvantage of only supplying intermittent power for electricity generation. This inconvenience can be removed through CSP technology, which together with a suitable heat storage system can generate electricity even with cloudy skies or after sunset. Thermal Energy Storage allows the mitigation of short fluctuations and extension of electricity supply to more desirable periods, making Concentrated Solar Power dispatchable [74].

This work outlines an analysis of the latest scientific literature in order to achieve the state-of-the-art review of the development and worldwide implementation of CSP and TES technologies. In this respect, the following aspects were presented: CSP's position in worldwide policies and targets for Renewable Energy Sources (RES) development, global distribution DNI map alongside to the global map of suitability for installation of solar thermal plants, the concept and the main CSP technologies together with the main TES methods, and, in the end, there have been discussions about the worldwide stage installation of capacities based on CSP. The conclusions can be summarized as follows:

- CSP technology along with TES maximizes the potential of solar energy through polygeneration, providing the opportunity to obtain more energy forms used by a unique resource.
- The main specific characteristic of CSP technologies compared to other renewable energy conversion equipment is the use of a heat storage system to generate electricity even when cloudy or after the sunset due to DNI, in contrast to the photovoltaic panels (PV).
- A percentage of 85% of all CSP research projects is carried out on the parabolic troughs (PT) technology, according to the data available on operating experience.
- Sensible heat storage technology is the most used in CSP plants in operation.
- LCOE estimates for CSP are still relatively high, but these technologies are in constant research and development, and as a number of pilot projects currently underway in this field will be validated and implemented on a large scale, including energy storage solutions, and they are expected to result in declining costs in the near future.
- In terms of power generated by CSP capacity, Spain ranks first in the world ranking, being followed by United States, which ranks second, and then South Africa, India, Morocco, China.
- From the statistical data presented by U.S. National Renewable Energy Laboratory (NREL), a significant increase in CSP capacity from 0.4 gigawatts produced in 2007, to 4.9 gigawatts in 2017, can be noticed.
- Several countries, in addition to the different policies for solar power generation, with which they face the appropriate system planning and operations for power supply systems to provide reliable quality and electrical power, make use of new eco-sustainable plants systems to reduce pollutant emissions and energy consumption [75–77].

One R&D project allowed ICSI Râmnicu Vâlcea to purchase an integrated CSP Dish Stirling SUNFLOWER 35 type system for polygeneration of energy by concentrating solar irradiation, which has been installed in the current location, together with all the accessories necessary for good operation in optimal conditions. The Stirling engine power unit was installed within the CSP Dish Stirling SUNFLOWER 35 integrated system, but only after this equipment had been pre-tested and put into operation.

Author Contributions: Conceptualization, G.B. and C.F.; Methodology, R.-A.F., M.S.R.; Software, A.L.; Validation, G.B., R.-A.F. and M.S.R.; Formal Analysis, M.R.; Investigation, G.R. and A.E.; Resources, C.F.; Data Curation, C.F.; Writing—Original Draft Preparation, R.-A.F., G.R.; Writing—Review & Editing, M.S.R.; Visualization, A.L.; Supervision, G.B., C.F.; Project Administration, C.F.; Funding Acquisition, M.S.R.

Funding: This research received no external funding.

Acknowledgments: This work was supported by a grant of the Romanian Ministry of Research and Innovation, CCCDI-UEFISCDI, project number PN-III-P1-1.2-PCCDI-2017-0776/No. 36 PCCDI/15.03.2018, within PNCDI III.

Conflicts of Interest: "The authors declare no conflict of interest."

Nomenclature

Abbreviation	Definition
CO_2	Carbon dioxide
CSP	Concentrated Solar Power

DNI	Direct Normal Irradiance
EC	European Commission
EU	European Union
ISO	International Organization for Standardization
LCOE	Levelized Cost of Electricity
LFR	Linear Fresnel Reflector
MENA	Middle East and North Africa
PD	Parabolic Dish
PT	Parabolic Trough
PV	Photovoltaic Panel
RES	Renewable Energy Source
STE	Solar Thermal Energy
TES	Thermal Energy Storage
USA	United State of America
WMO	World Meteorological Organization

References

1. Bejan, M.; Rusu, T.; Bălan, I. The Present and Future of Energy, Part. II, Renewable Energies. Technological perspectives. *Buletinul AGIR* **2010**, *4*, 178–183.
2. Owusu, P.A.; Asumadu-Sarkodie, S. A review of renewable energy sources, sustainability issues and climate change mitigation. *Cogent Eng.* **2016**, *3*, 1167990. [CrossRef]
3. European Commission. *Directive 2009/28/EC on the Promotion of the Use of Energy from Renewable Sources*; European Parliament, Council of the European Union: Brussels, Belgium, 2009.
4. *Renewable Energy Policy Network for the 21st Century (REN21)*; Renewables 2017 Global Status Report; REN21: Paris, France, September 2017; ISBN 978-3-9818107-6-9.
5. Burke, M.J.; Stephens, J.C. Political power and renewable energy futures: A critical review. *Energy Res. Soc. Sci.* **2018**, *35*, 78–93. [CrossRef]
6. Kabir, E.; Kumar, P.; Kumar, S.; Adelodun, A.A.; Kim, K.H. Solar energy: Potential and future prospects. *Renew. Sustain. Energy Rev.* **2018**, *82*, 894–900. [CrossRef]
7. Kannan, N.; Vakeesan, D. Solar energy for future world: A review. *Renew. Sustain. Energy Rev.* **2016**, *62*, 1092–1105. [CrossRef]
8. Strantzali, E.; Aravossis, K. Decision making in renewable energy investments: A review. *Renew. Sustain. Energy Rev.* **2016**, *55*, 885–898. [CrossRef]
9. European Parliament. *Energia din Surse Regenerabile; Fise Tehnice UE—2016*; European Parliament: Brussels, Belgium, 2016.
10. Felseghi, R.A.; Papp, A. Considerations regarding hybrid systems of power generator from renewable energy sources, Studia Universitatis Babes-Bolyai. *Ambientum* **2014**, *59*, 27–40.
11. González-Roubaud, E.; Pérez-Osorio, D.; Prieto, C. Review of commercial thermal energy storage in concentrated solar power plants: Steam vs. molten salts. *Renew. Sustain. Energy Rev.* **2017**, *80*, 133–148. [CrossRef]
12. International Energy Agency (IEA); International Renewable Energy Agency (IRENA). *Concentrating Solar Power: Technology Brief*; IEA & IRENA: Abu Dhabi, UAE, 2013.
13. Balghouthi, M.; Trabelsi, S.E. Mahmoud Ben Amara, Abdessalem Bel Hadj Ali, Guizani, A. Potential of concentrating solar power (CSP) technology in Tunisia and the possibility of interconnection with Europe. *Renew. Sustain. Energy Rev.* **2016**, *56*, 1227–1248. [CrossRef]
14. Cannistraro, G.; Cannistraro, M.; Trovato, G. Island "Smart Energy" for Eco-Sustainable Energy—A Case Study "Favignana Island". *Int. J. Heat Techol.* **2017**, *35*. [CrossRef]
15. European Commission. *Directive of the European Parliament and of the Council on the Promotion of the Use of Energy from Renewable Sources*; European Commission: Brussels, Belgium, 2016; p. 18.
16. Corsiuc, G.; Marza, C.; Ceuca, E.; Felseghi, R.A.; Şoimoşan, T.M. Hybrid Solar-Wind Stand-Alone Energy System: A Case Study. *Appl. Mech. Mater.* **2015**, *772*, 536–540. [CrossRef]

17. National Renewable Energy Laboratory (NREL). *Renewable Electricity Futures Study, Vol. 2. Renewable Electricity Generation and Storage Technologies*; NREL/TP-6A20-52409-2; National Renewable Energy Laboratory (NREL): Golden, CO, USA, 2012.
18. Hosenuzzaman, M.; Rahim, N.A.; Selvaraj, J.; Hasanuzzaman, M.; Malek, A.B.M.A.; Nahar, A. Global prospects, progress, policies, and environmental impact of solar photovoltaic power generation. *Renew. Sustain. Energy Rev.* **2015**, *41*, 284–297. [CrossRef]
19. *Renewables 2017. Global Status Report*; REN21: Paris, France, 2017; p. 23.
20. *Renewables 2017. Global Status Report*; REN21: Paris, France, 2017; p. 25.
21. Şoimoşan, T.M.; Felseghi, R.A. Influence of the thermal energy storage on the hybrid heating systems' energy profile. *Prog. Cryog. Isotopes Sep.* **2016**, *19*, 5–18.
22. Aschilean, I.; Varlam, M.; Culcer, M.; Iliescu, M.; Raceanu, M.; Enache, A.; Raboaca, M.S.; Rasoi, G.; Filote, C. Hybrid Electric Powertrain with Fuel Cells for a Series Vehicle. *Energies* **2018**, *11*, 1294. [CrossRef]
23. Abdmouleh, Z.; Gastli, A.; Brahim, L.B.; Haouari, M.; Al-Emadi, N.A. Review of optimization techniques applied for the integration of distributed generation from renewable energy sources. *Renew. Energy* **2017**, *113*, 266–280. [CrossRef]
24. Khan, J.; Arsalan, M.H. Solar power technologies for sustainable electricity generation—A review. *Renew. Sustain. Energy Rev.* **2016**, *55*, 414–425. [CrossRef]
25. Pandey, A.K.; Tyagi, V.V.; Selvaraj, J.A.L.; Rahim, N.A.; Tyagi, S.K. Recent advances in solar photovoltaic systems for emerging trends and advanced applications. *Renew. Sustain. Energy Rev.* **2016**, *53*, 859–884. [CrossRef]
26. Carson, L. *Australian Energy Resource Assessment*, 2nd ed.; Report, Cap. 10—Solar Energy; Geoscience Australia: Canberra, Australia, 2014.
27. Pentiuc, R.D.; Popa, C.D.; Dascalu, A. The Influence of LED Street Lighting Upon Power Quality in Electrical Networks. In Proceedings of the 8th International Conference and Exposition on Electrical and Power Engineering (EPE) Romania, Iasi, Romania, 16–18 October 2014. [CrossRef]
28. Aschilean, I.; Rasoi, G.; Raboaca, M.S.; Filote, C.; Culcer, M. Design and Concept of an Energy System Based on Renewable Sources for Greenhouse Sustainable Agriculture. *Energies* **2018**, *11*, 1201. [CrossRef]
29. Marza, C.; Corsiuc, G.D.; Pop, A. Case study regarding the efficiency of electricity generation using photovoltaic panels. In Proceeding of the International Multidisciplinary Scientific GeoConference SGEM 2018, Varna, Bulgaria, 30 June–9 July 2018. Section Renewable Energy Sources and Clean Technologies.
30. Edenhofer, O.; Pichs Madruga, R.; Sokona, Y.; Seyboth, K.; Matschoss, P.; Kadner, S.; Zwickel, T.; Eickemeier, P.; Hansen, G.; Schloemer, S.; et al. *Renewable Energy Sources and Climate Change Mitigation*; Special Report of the Intergovernmental Panel on Climate Change; Cambridge University Press: Cambridge, UK, 2012; pp. 607–614.
31. Liu, M.; Tay, N.H.S.; Bell, S.; Belusko, M.; Jacob, R.; Will, G.; Saman, W.; Bruno, F. Review on concentrating solar power plants and new developments in high temperature thermal energy storage technologies. *Renew. Sustain. Energy Rev.* **2016**, *53*, 1411–1432. [CrossRef]
32. Blanc, P.; Espinar, B.; Geuder, N.; Gueymard, C.; Meyer, R.; Pitz-Paal, R.; Reinhardt, B.; Renne, D.; Sengupta, M.; Wald, L.; et al. Direct normal irradiance related definitions and applications: The circumsolar issue. *Sol. Energy* **2014**, *110*, 561–577. [CrossRef]
33. Law, E.W.; Kay, M.; Taylor, R.A. Evaluating the benefits of using short-term direct normal irradiance forecasts to operate a concentrated solar thermal plant. *Sol. Energy* **2016**, *140*, 93–108. [CrossRef]
34. WMO. *CIMO Guide to Meteorological Instruments and Methods of Observation*, 7th ed.; Measurement of Radiation, WMO-No. 8 (2010 Update); World Meteorological Organization: Geneva, Switzerland, 2010; Chapter 7; pp. 157–198.
35. Gueymard, C.A.; Ruiz-Arias, J.A. Extensive worldwide validation and climate sensitivity analysis of direct irradiance predictions from 1-min global irradiance. *Sol. Energy* **2016**, *128*, 1–30. [CrossRef]
36. Direct Normal Irradiation. Available online: https://solargis.com/maps-and-gis-data/download/world (accessed on 25 October 2018).
37. Shouman, E.R.; Hesham, E. Forecasting transition electricity solar energy from MENA to Europe. *ARPN J. Eng. Appl. Sci.* **2016**, *11*, 3029–3040.
38. Law, E.W.; Kay, M.; Taylor, R.A. Calculating the financial value of a concentrated solar thermal plant operated using direct normal irradiance forecasts. *Sol. Energy* **2016**, *125*, 267–281. [CrossRef]

39. Vieira de Souza, L.E.; Gilmanova Cavalcante, A.M. Concentrated Solar Power deployment in emerging economies: The cases of China and Brazil. *Renew. Sustain. Energy Rev.* **2017**, *72*, 1094–1103. [CrossRef]

40. Lovegrove, K.; Wyder, J.; Agrawal, A.; Borhuah, D.; McDonald, J.; Urkalan, K. *Concentrating Solar Power in India*; Report Commissioned by the Australian Government and Prepared by ITP Power; ITP Power: Bangkok, Thailand, 2011; ISBN 978-1-921299-52-0.

41. Jiménez-Torres, M.; Rus-Casas, C.; Lemus-Zúiga, L.G.; Hontoria, L. The Importance of Accurate Solar Data for Designing Solar Photovoltaic Systems—Case Studies in Spain. *Sustainability* **2017**, *9*, 247. [CrossRef]

42. Available online: https://sibved.livejournal.com/171584.html (accessed on 12 June 2018).

43. Bloess, A.; Schill, W.P.; Zerrahn, A. Power-to-heat for renewable energy integration: A review of technologies, modeling approaches, and flexibility potentials. *Appl. Energy* **2018**, *212*, 1611–1626. [CrossRef]

44. CSP Today Global Tracker. Available online: http://social.csptoday.com/tracker/projects (accessed on 15 June 2018).

45. Guenounou, A.; Malek, A.; Aillerie, M. Comparative performance of PV panels of different technologies over one year of exposure: Application to a coastal Mediterranean region of Algeria. *Energy Convers. Manag.* **2016**, *114*, 356–363. [CrossRef]

46. Available online: http://analysis.newenergyupdate.com/csp-today/technology/trouble-dish-stirling-csp (accessed on 12 June 2018).

47. Breyer, C.; Bogdanov, D.; Aghahosseini, A.; Gulagi, A.; Child, M.; Oyewo, A.S.; Farfan, J.; Sadovskaia, K.; Vainikka, P. Solar photovoltaics demand for the global energy transition in the power sector. *Wiley Prog. Photovolt. Res. Appl.* **2018**, *26*, 505–523. [CrossRef]

48. Hu, A.H.; Huang, L.H.; Lou, S.; Kuo, C.H.; Huang, C.Y.; Chian, K.J.; Chien, H.T.; Hong, H.F. Assessment of the Carbon Footprint, Social Benefit of Carbon Reduction, and Energy Payback Time of a High-Concentration Photovoltaic System. *Sustainability* **2017**, *9*, 27. [CrossRef]

49. International Energy Agency (IEA). *Technology Roadmap, Solar Thermal Electricity*; International Energy Agency (IEA): Paris, France, 2016.

50. International Energy Agency (IEA). *Technology Roadmap, Solar Thermal Electricity*; International Energy Agency (IEA): Paris, France, 2010.

51. International Energy Agency (IEA). *Technology Roadmap, Solar Thermal Electricity*; International Energy Agency (IEA): Paris, France, September 2014.

52. Powell, K.M.; Rashid, K.; Ellingwood, K.; Tuttle, J.; Iverson, B.D. Hybrid concentrated solar thermal power systems: A review. *Renew. Sustain. Energy Rev.* **2017**, *80*, 215–237. [CrossRef]

53. Available online: https://www.iea.org/newsroom/news/2017/january/making-freshwater-from-the-sun.html (accessed on 25 October 2018).

54. Pelay, U.; Luo, L.; Fan, Y.; Stitou, D.; Rood, M. Thermal energy storage systems for concentrated solar power plants. *Renew. Sustain. Energy Rev.* **2017**, *79*, 82–100. [CrossRef]

55. Dowling, A.W.; Zheng, T.; Zavala, V.M. Economic assessment of concentrated solar power technologies: A review. *Renew. Sustain. Energy Rev.* **2017**, *72*, 1019–1032. [CrossRef]

56. Giurca, I.; Aşchilean, I.; Naghiu, G.S.; Badea, G. Selecting the Technical Solutions for Thermal and Energy Rehabilitation and Modernization of Buildings. *Procedia Technol.* **2016**, *22*, 789–796. [CrossRef]

57. Gaglia, A.G.; Lykoudis, S.; Argiriou, A.A.; Balaras, C.A.; Dialynas, E. Energy efficiency of PV panels under real outdoor conditionse. An experimental assessment in Athens, Greece. *Renew. Energy* **2017**, *101*, 236–243. [CrossRef]

58. Pentiuc, R.D.; Vlad, V.; Lucache, D.D. Street Lighting Power Quality. In Proceedings of the 8th International Conference and Exposition on Electrical and Power Engineering (EPE), Iasi, Romania, 16–18 October 2014. [CrossRef]

59. Gábor Pintér, G.; Hegedüsné Baranyai, N.; Wiliams, A.; Zsiborács, H. Study of Photovoltaics and LED Energy Efficiency: Case Study in Hungary. *Energies* **2018**, *11*, 790. [CrossRef]

60. Naghiu, G.S.; Giurca, I.; Aşchilean, I.; Badea, G. Multicriterial analysis on selecting solar radiation concentration ration for photovoltaic panels using Electre-Boldur method. *Procedia Technol.* **2016**, *22*, 773–780. [CrossRef]

61. Corona, B.; Ruiz, D.; San Miguel, G. Life Cycle Assessment of a HYSOL Concentrated Solar Power Plant: Analyzing the Effect of Geographic Location. *Energies* **2016**, *9*, 413. [CrossRef]

62. Kumara, A.; Sahb, B.; Singh, A.R.; Deng, Y.; He, X.; Kumar, P.; Bansal, R.C. A review of multi criteria decision making (MCDM) towards sustainable renewable energy development. *Renew. Sustain. Energy Rev.* **2017**, *69*, 596–609. [CrossRef]

63. Salvatore, J. World Energy Perspective—Cost of Energy Technologies. Bloomberg New Energy Finance, London. 2013. Available online: https://www.worldenergy.org/wp-content/uploads/2013/09/WEC_J1143_CostofTECHNOLOGIES_021013_WEB_Final.pdf (accessed on 17 October 2018).

64. Ehtiwesh, I.A.S.; Coelho, M.C.; Sousa, A.C.M. Exergetic and environmental life cycle assessment analysis of concentrated solar power plants. *Renew. Sustain. Energy Rev.* **2016**, *56*, 145–155. [CrossRef]

65. Shravanth Vasisht, M.; Srinivasan, J.; Ramasesha, S.K. Performance of solar photovoltaic installations: Effect of seasonal variations. *Sol. Energy* **2016**, *131*, 39–46. [CrossRef]

66. Zsiborács, H.; Hegedüsné Baranyai, N.; Vincze, A.; Háber, I.; Pintér, G. Economic and Technical Aspects of Flexible Storage Photovoltaic Systems in Europe. *Energies* **2018**, *11*, 1445. [CrossRef]

67. International Renewable Energy Agency (IRENA). Renewable Power Generation Costs. Available online: https://www.irena.org/-/media/Files/IRENA/Agency/Publication/2018/Jan/IRENA_2017_Power_Costs_2018.pdf (accessed on 20 June 2018).

68. Toro, C.; Rocco, M.V.; Colombo, E. Exergy and Thermoeconomic Analyses of Central Receiver Concentrated Solar Plants Using Air as Heat Transfer Fluid. *Energies* **2016**, *9*, 885. [CrossRef]

69. Akbari, H.; Browne, M.C.; Ortega, A.; Huang, M.J.; Hewitt, N.J.; Norton, B.; McCormack, S.J. Efficient energy storage technologies for photovoltaic systems. *Sol. Energy* **2018**, in press. [CrossRef]

70. Renewable Energy Policy Network for 21st Century REN21. Global Status Report Renewables 2018. Available online: http://www.ren21.net/wp-content/uploads/2018/06/17-8652_GSR2018_FullReport_web_final_.pdf (accessed on 28 October 2018).

71. National Renewable Energy Laboratory (NREL) Concentrating Solar Power Projects. Available online: https://solarpaces.nrel.gov/by-status (accessed on 30 October 2018).

72. National Renewable Energy Laboratory (NREL) Project Listing. Available online: http://www.nrel.gov/csp/solarpaces/ (accessed on 30 October 2018).

73. CSP World Website. Available online: http://www.cspworld.org/cspworldmap (accessed on 5 January 2018).

74. Pentiuc, R.D.; Baluta, G.; Popa, C.; Mahalu, G. Calculation of Reactances for Ring Windings to Toroidal Inductors of Hybrid Induction Machine. *Adv. Electr. Comput. Eng.* **2009**, *9*, 70–74. [CrossRef]

75. Cannistraro, M.; Mainardi, E.; Bottarelli, M. Testing a Dual-Source Heat Pump. *Math. Model. Eng. Probl.* **2018**, *5*, 205–221. [CrossRef]

76. Cannistraro, M.; Castelluccio, M.E.; Germanò, D. New Sol-Gel Deposition Techinique in the Smart-Windows Computation of Possible Applications of Smart-Windows in Buildings. *J. Build. Eng.* **2018**, *19*, 295–301. [CrossRef]

77. Cannistraro, M.; Chao, J.; Ponterio, L. Experimental study of air pollution in the urban centre of the city of Messina. *Model. Meas. Control C* **2018**, *79*, 133–139. [CrossRef]

Article

Aggregation of Type-4 Large Wind Farms Based on Admittance Model Order Reduction

Jaime Martínez-Turégano *, Salvador Añó-Villalba, Soledad Bernal-Perez and Ramon Blasco-Gimenez

Instituto de Automática e Informática Industrial, Universitat Politècnica de València, 46022 Valencia, Spain; sanyo@die.upv.es (S.A.-V.); sbernal@die.upv.es (S.B.-P.); r.blasco@ieee.org (R.B.-G.)
* Correspondence: jaumartu@upv.es

Received: 10 April 2019; Accepted: 30 April 2019; Published: 7 May 2019

Abstract: This paper presents an aggregation technique based on the resolution of a multi-objective optimization problem applied to the admittance model of a wind power plant (WPP). The purpose of the presented aggregation technique is to reduce the order of the wind power plant model in order to accelerate WPP simulation while keeping a very similar control performance for both the simplified and the detailed models. The proposed aggregation technique, based on the admittance model order reduction, ensures the same DC gain, the same gain at the operating band frequency, and the same resonant peak frequency as the detailed admittance model. The proposed aggregation method is validated considering three 400-MW grid-forming Type-4 WPPs connected to a diode rectifier HVDC link. The proposed aggregation technique is compared to two existing aggregation techniques, both in terms of frequency and time response. The detailed and aggregated models have been tested using PSCAD-EMTsimulations, with the proposed aggregated model leading to a 350-fold reduction of the simulation time with respect to the detailed model. Moreover, for the considered scenario, the proposed aggregation technique offers simulation errors that are, at least, three-times smaller than previously-published aggregation techniques.

Keywords: off-shore wind farms; wind farm aggregation; admittance model order reduction; HVDC diode rectifiers; grid-forming wind turbines

---

1. Introduction

Large off-shore wind power plants (WPPs) with more than one hundred wind turbine generators (WTG) are currently operational or in the planning or approval phases. Simulation studies covering individual WTGs during the design phase are very demanding from the computational point of view. When studying the impact of WPPs in the overall HVDC or HVAC transmission network, very detailed models are not generally required. Therefore, simulation complexity and computing time can be reduced by using aggregated WTG models. However, these aggregated models should provide a faithful representation of the actual WPP dynamics.

This is particularly important if the WPP consists of grid-forming converters, such as when using diode rectifier (DR)-based HVDC stations for the connection of the offshore WPP. The use of DR HVDC stations can significantly reduce capital expenditure (CAPEX) and operational expenditure (OPEX) by increasing the efficiency and robustness of the overall system [1–6]. However, a large number of case studies is required in order to verify the correct integration of DR-connected WPPs.

Therefore, this paper is focused on the procedure to obtain an aggregated WTG model for large Type-4 WPPs that allows the verification of grid-forming WTG controllers for DR HVDC station connection. The presented aggregation technique aims at reducing the computational requirements of such a verification, while achieving better simulation accuracy than current state-of-the-art aggregation techniques.

A large amount of literature has been developed on aggregation techniques for different applications [7,8]. For example, the work in [9] presented an aggregation technique based on power losses, while [10] presented an aggregation technique for grid disturbances studies. The work in [11] included a method for Type-3 and Type-1 WTG aggregation suitable for large power systems simulation. In this case, the WPP grid was aggregated using a short-circuit impedance method. In [12], an aggregated method based on voltage drop assumptions was performed as the most appropriate aggregation method for stability assessment studies. On the other hand, in [13], an aggregation method was proposed in order to represent the whole power system seen from the WTG AC terminals. Finally, an improvement on the aggregation method based on power losses for off-shore WPP with a diode-based HVDC link was proposed in [14].

However, none of the previous methods were based on a model order reduction of the total WPP admittance model. The proposed model order reduction leads to better matching between the dynamic behavior of the aggregated and full WPP models, as the aggregated system keeps the original DC admittance, fundamental frequency admittance, and first resonance peak characteristics of the original system.

The paper is organized as follows. The second section shows the procedure to obtain the full admittance model of a typical WPP, consisting of a number of radially-connected WTG strings, as this is the typical topology used in WPPs. Other WPP topologies (i.e., with different degrees of meshing) can also be aggregated with the proposed method, by suitable modification of the analytical admittance calculation. The third section explains how to apply the proposed aggregation technique. The fourth section describes the system that is used in the fifth section to compare the proposed aggregation technique with other existing aggregation techniques found in the literature. The sixth section compares the aggregated and detailed model of a WPP using PSCADsimulations, considering three different cases. The last section includes the discussion of the proposed aggregation technique compared with other existing aggregation algorithms.

2. Full-Order Admittance System Modeling

The proposed model aggregation technique was based on the resolution of a multi-objective optimization problem applied to a full-WPP admittance model. This section includes the analytical development of the WPP admittance model. The inputs to the model were the individual WTG voltages, and its output was the WPP current at the point of common coupling (PCC).

Each considered WPP consisted of n radial-connected strings, with each string consisting of m_i cascaded connected WTGs. This topology corresponds to the great majority of existing WPPs, as ring-connected strings are always exploited radially. In any case, it is very easy to modify the proposed analytical model to calculate the overall grid-admittance of other WPP topologies.

Figure 1 shows the WTG-j ($j = 1, 2, ..., m_i$) in the string-i ($i = 1, 2, ..., n$) of the WPP. U_{ij} is the WTG grid-side converter, and (L_{Tij} and R_{Tij}) represents the WTG transformer. Several alternatives have been considered to model the array cables, e.g., [15] added parallel L-R branches to a PI-model in order to represent the frequency-dependent impedance of the cable. This model has been used in [16,17] to obtain a state-space model and reduce the order of the cable model. Additionally, a frequency-dependent PI model of a three-core submarine cable was studied in [18].

In the presented case study, PI-sections will be used to represent the cable dynamics (C_{Lij}, L_{Lij}, and R_{Lij}), since the cable length between WTGs was less than 3 km with its resonant peak at frequencies higher than 1000 Hz. The use of PI-models offers a good trade-off between complexity and accuracy.

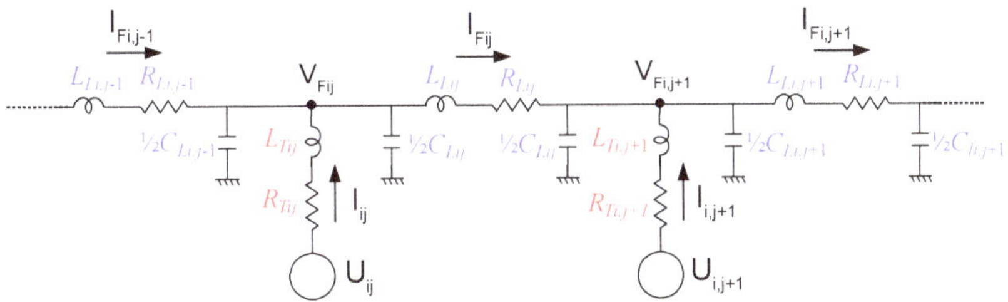

Figure 1. WTG-j ($j = 1, 2, ..., m_i$) in the string-i ($i = 1, 2, ..., n$) of the wind power plant (WPP).

2.1. String Admittance Model

The dynamics of the WTG-j ($j = 1, 2, ..., m_i$) and cable shown in Figure 1 can be written as:

$$U_{ij} = R_{Tij} \cdot I_{ij} + L_{Tij} \cdot \frac{dI_{ij}}{dt} + V_{Fij} \tag{1}$$

$$I_{Fi,j-1} + I_{ij} - I_{Fij} = \left(\frac{C_{Li,j-1}}{2} + \frac{C_{Lij}}{2} \right) \cdot \frac{dV_{Fij}}{dt} \tag{2}$$

$$V_{Fij} = R_{Lij} \cdot I_{Fij} + L_{Lij} \cdot \frac{dI_{Fij}}{dt} + V_{Fi,j+1} \tag{3}$$

where I_{ij} is the current from WTG-j, V_{Fij} is the voltage at the secondary side of the transformer, and I_{Fij} is the current through the cable. Note that for $j = 1$ (leftmost WTG), the current $I_{Fi,j-1}$ did not exist, and for $j = m_i$ (rightmost WTG), the voltage $V_{Fi,j+1}$ was the PCC voltage (V_F), which was considered a disturbance.

From Equations (1)–(3), the state-space admittance model of string-i ($i = 1, 2, ..., n$) can be calculated as:

$$\dot{x}_{st,i} = A_{st,i} \cdot x_{st,i} + B_{st,i} \cdot u_{st,i} + W_{st,i} \cdot V_F \tag{4}$$

$$y_{st,i} = C_{st,i} \cdot x_{st,i} \tag{5}$$

where the state vector $x_{st,i}$ and the input vector $u_{st,i}$ are:

$$x_{st,i} = \begin{bmatrix} I_{i1} & V_{Fi1} & I_{Fi1} & I_{i2} & V_{Fi2} & I_{Fi2} & \cdots & I_{i,m_i} & V_{Fi,m_i} & I_{Fi,m_i} \end{bmatrix}, \quad u_{st,i} = \begin{bmatrix} U_{i1} & U_{i2} & \cdots & U_{i,m_i} \end{bmatrix}^T$$

Matrices $A_{st,i}$, $B_{st,i}$, and $C_{st,i}$ are the state, input, and output matrices, respectively:

$$A_{st,i} = \begin{bmatrix}
\frac{-R_{Ti1}}{L_{Ti1}} & \frac{-1}{L_{Ti1}} & 0 & 0 & 0 & 0 & & 0 & 0 & 0 \\
\frac{2}{C_{Li1}} & 0 & \frac{-2}{C_{Li1}} & 0 & 0 & 0 & & 0 & 0 & 0 \\
0 & \frac{1}{L_{Li1}} & \frac{-R_{Li1}}{L_{Li1}} & 0 & \frac{-1}{L_{Li1}} & 0 & & 0 & 0 & 0 \\
0 & 0 & 0 & \frac{-R_{Ti2}}{L_{Ti2}} & \frac{-1}{L_{Ti2}} & 0 & & 0 & 0 & 0 \\
0 & 0 & \frac{2}{C_{Li1}+C_{Li2}} & \frac{2}{C_{Li1}+C_{Li2}} & 0 & \frac{-2}{C_{Li1}+C_{Li2}} & & 0 & 0 & 0 \\
0 & 0 & 0 & 0 & \frac{1}{L_{Li2}} & \frac{-R_{Li2}}{L_{Li2}} & & 0 & 0 & 0 \\
 & & & & & & \ddots & & & \\
0 & 0 & 0 & 0 & 0 & 0 & & \frac{-R_{Ti,m_i}}{L_{Ti,m_i}} & \frac{-1}{L_{Ti,m_i}} & 0 \\
0 & 0 & 0 & 0 & 0 & 0 & & \frac{2}{C_{Li,m_i-1}+C_{Li,m_i}} & 0 & \frac{-2}{C_{Li,m_i-1}+C_{Li,m_i}} \\
0 & 0 & 0 & 0 & 0 & 0 & & 0 & \frac{1}{L_{Li,m_i}} & \frac{-R_{Li,m_i}}{L_{Li,m_i}}
\end{bmatrix}$$

$$B_{st,i} = \begin{bmatrix} \frac{1}{L_{Ti1}} & 0 & 0 \\ 0 & 0 & 0 \\ 0 & 0 & 0 \\ 0 & \frac{1}{L_{Ti2}} & 0 \\ 0 & 0 & 0 \\ 0 & 0 & 0 \\ & & \ddots \\ 0 & 0 & \frac{1}{L_{Ti,m_i}} \\ 0 & 0 & 0 \\ 0 & 0 & 0 \end{bmatrix}, \quad C_{st,i} = \begin{bmatrix} 0 & 0 & 0 & 0 & 0 & 0 & \cdots & 0 & 0 & 1 \end{bmatrix}$$

and:

$$W_{st,i} = \begin{bmatrix} 0 & 0 & 0 & 0 & 0 & 0 & \cdots & 0 & 0 & \frac{-1}{L_{Li,m_i}} \end{bmatrix}^T$$

2.2. Overall Admittance Model

The admittance model of the complete WPP was obtained by the radial connection of the n strings to the PCC (represented by voltage V_F). The state space dynamics of the overall system are:

$$\dot{x}_o = A_o \cdot x_o + B_o \cdot u_o + W_o \cdot V_F \tag{6}$$

$$y_o = C_o \cdot x_o \tag{7}$$

where A_o, B_o, and C_o are the state, input, and output of the overall system obtained by combining the string state space equations:

$$A_o = \begin{bmatrix} A_{st1} & 0 & 0 \\ 0 & A_{st2} & 0 \\ & & \ddots \\ 0 & 0 & A_{st,n} \end{bmatrix}, \quad B_o = \begin{bmatrix} B_{st1} & 0 & 0 \\ 0 & B_{st2} & 0 \\ & & \ddots \\ 0 & 0 & B_{st,n} \end{bmatrix}$$

$$C_o = \begin{bmatrix} C_{st1} & C_{st2} & \cdots & C_{st,n} \end{bmatrix}$$

and:

$$W_o = \begin{bmatrix} W_{st1} & W_{st2} & \cdots & W_{st,n} \end{bmatrix}^T$$

3. Aggregation Technique

The structure of the aggregated equivalent WPP is shown in Figure 2. At this point, the aggregation technique based on multi-objective optimization was carried out by following these steps:

- Input reduction.
- Multi-objective optimization problem statement.
- Multi-objective optimization problem solution.

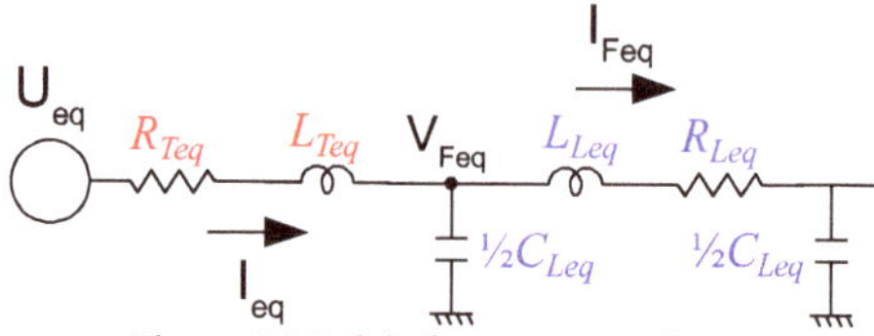

Figure 2. Model of an aggregated WPP.

3.1. Input Reduction

The proposed admittance model had a total of n voltage inputs that should be reduced to a single voltage input. Using the admittance model of the overall system, the superposition principle can be applied in order to obtain a single-input single-output model. The equivalent single-input voltage source was obtained by introducing the vector σ in Equations (6) and (7), leading to Equations (8) and (9):

$$\dot{x} = A \cdot x + B \cdot \sigma \cdot u \tag{8}$$

$$y = C \cdot x + D \cdot \sigma \cdot u \tag{9}$$

σ is a vector with values of one for connected WTGs or zero for disconnected WTG:

$$\sigma = \left(\sigma_{1,1}, \sigma_{1,2}, \cdots, \sigma_{i,j}, \cdots, \sigma_{n,m_i} \right)^T$$

The obtained SISO system had a single equivalent voltage input (u) and the WPP PCC current as an output ($I_F = I_{F1,m_1} + I_{F2,m_2} + \dots + I_{Fn,m_n}$) and had a total of $3 \cdot n$ states, with n equal to the number of WTGs in the WPP.

3.2. Multi-Objective Optimization Problem Statement

The target aggregated model is shown in Figure 2, with dynamics:

$$\dot{x}_{eq} = A_{eq} \cdot x_{eq} + B_{eq} \cdot u_{eq} \tag{10}$$

$$y_{eq} = C_{eq} \cdot x_{eq} + D_{eq} \cdot u_{eq} \tag{11}$$

where the inputs of the aggregated WTG model u_{eq} were the aggregated voltage source (U_{eq}) and the voltage at the point of common coupling (V_F). The model output was the WPP current (I_{Feq}), and the state variables were I_{eq}, V_{Feq}, and I_{Feq}, as shown in Figure 2.

Matrices A_{eq}, B_{eq}, C_{eq}, and D_{eq} are given below:

$$A_{eq} = \begin{pmatrix} -\dfrac{R_{Teq}}{L_{Teq}} & \dfrac{-1}{L_{Teq}} & 0 \\[2ex] \dfrac{1}{\frac{C_{Leq}}{2}} & 0 & \dfrac{-1}{\frac{C_{Leq}}{2}} \\[2ex] 0 & \dfrac{1}{L_{Leq}} & -\dfrac{R_{Leq}}{L_{Leq}} \end{pmatrix} \qquad B_{eq} = \begin{pmatrix} \dfrac{1}{L_{Teq}} & 0 \\[2ex] 0 & 0 \\[2ex] 0 & -\dfrac{1}{L_{Leq}} \end{pmatrix}$$

$$C_{eq} = \begin{pmatrix} 0 & 0 & 1 \end{pmatrix} \qquad D_{eq} = \begin{pmatrix} 0 & 0 \end{pmatrix}$$

The reduced order model in Equations (10) and (11) had five parameters to be identified (L_{Teq}, R_{Teq}, L_{Leq}, R_{Leq}, and C_{Leq}). In order to formulate the multi-optimization problem properly, the following considerations were made:

- The aggregated capacitance C_{Leq} was considered as the sum of the total shunt capacitance in the WPP grid [9].
- The factor X_T/R_T of the WTG transformer shall be maintained in the aggregated transformer ($X_{Ti}/R_{Ti} = X_{Teq}/R_{Teq}$).

At this point, three objectives are proposed to identify the remaining three parameters for the aggregated model, namely:

- Minimize the error between the DC gain of aggregated ($K_{\omega=0}$ Hz) and the high order SISO system ($K^*_{\omega=0}$ Hz).
- Minimize the error between the gain of the aggregated WTG model at the grid frequency ($K_{\omega=\omega_0}$ Hz) and that of the high order SISO system ($K^*_{\omega=\omega_0}$ Hz).
- Minimize the error between the frequency of the resonant peak (ω_{peak}) of the aggregated WTG model and the frequency of the first resonant peak of the high order SISO system (ω^*_{peak}).

Finally, we shall consider that the WTG model parameters to identify always shall be positive and within a defined range ($R_{Teq,min} < R_{Teq} < R_{Teq,max}$, $R_{Leq,min} < R_{Leq} < R_{Leq,max}$, and $L_{Leq,min} < L_{Leq} < L_{Leq,max}$).

From the assumptions above, the following multi-objective optimization function was defined for the aggregated model:

$$f(R_{Teq}, L_{Leq}, R_{Leq}) = \begin{pmatrix} \omega_{peak}(R_{Teq}, L_{Leq}, R_{Leq}) \\ K_{\omega=0}(R_{Teq}, L_{Leq}, R_{Leq}) \\ K_{\omega=\omega0}(R_{Teq}, L_{Leq}, R_{Leq}, \omega_0) \end{pmatrix} \tag{12}$$

The goal function (f^*) was obtained from the system described in Equations (8) and (9). Therefore, the aim of the optimization problem was to minimize the maximum of:

$$W_i(f(R_{Teq}, L_{Leq}, R_{Leq}) - f^*) \tag{13}$$

where W_i is a diagonal matrix containing the weights to scale each function of $f(R_{Teq}, L_{Leq}, R_{Leq})$. Additionally, the minimization problem shall satisfy the constraints:

$$\begin{pmatrix} R_{Teq,min} \\ L_{Leq,min} \\ R_{Leq,min} \end{pmatrix} < \begin{pmatrix} R_{Teq} \\ L_{Leq} \\ R_{Leq} \end{pmatrix} < \begin{pmatrix} R_{Teq,max} \\ L_{Leq,max} \\ R_{Leq,max} \end{pmatrix}$$

The multi-objective optimization problem was solved by the goal attainment method described in [19–22]. The criteria used to choose the initial parameter and weight estimates is included in Section 5.1.

4. HVDC Diode Rectifier-Connected Wind Power Plant

The system under study is shown in Figure 3. It consisted of three 400-MW WPPs connected to the on-shore network through a diode rectifier-based HVDC link [1,14].

Each WPP consisted of 50 Type-4 WTGs rated at 8 MW each. Each WTG generator had a full-scale back-to-back converter, a PWM filter, and a WTG transformer that were connected to the 66-kV off-shore AC grid.

Each one of the three WPPs ($k = 1, 2, 3$) was composed of 50 WTGs, distributed in $n = 7$ strings ($i = 1, 2, ..n$) with m_i wind turbines per string ($j = 1, 2, ..m_i$).

Finally, the rectifier station consisted of three diode rectifier platforms, DC-side series connected. The rectifier stations were connected by means of the HVDC cable to the onshore VSC station, as shown in Figure 3. Each DR station consisted of two 12-pulse DR bridges, together with the corresponding transformers and AC filters [14].

The implemented control is shown in Figure 4. The control consisted of a centralized controller for the total delivered active power P_{OWPP}:

$$P_{OWPP} = \sum_{k=1}^{3} \sum_{i=1}^{n} \sum_{j=1}^{m_i} P_{WTG,ijk}$$

and a distributed controller based on a standard P/ω, Q/V droop controller.

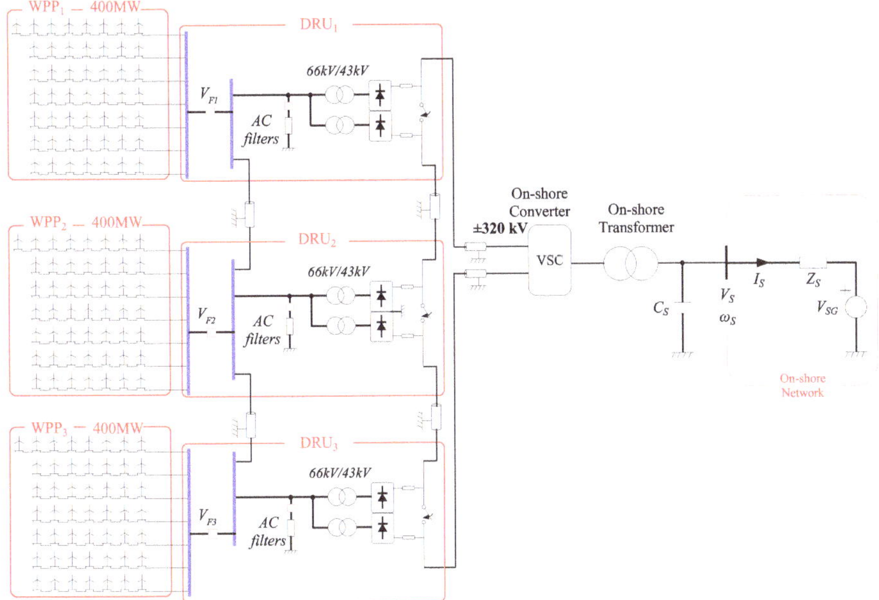

Figure 3. Off-shore WPP with three WPPs of 400 MW each connected to the on-shore grid through a diode-based HVDC link.

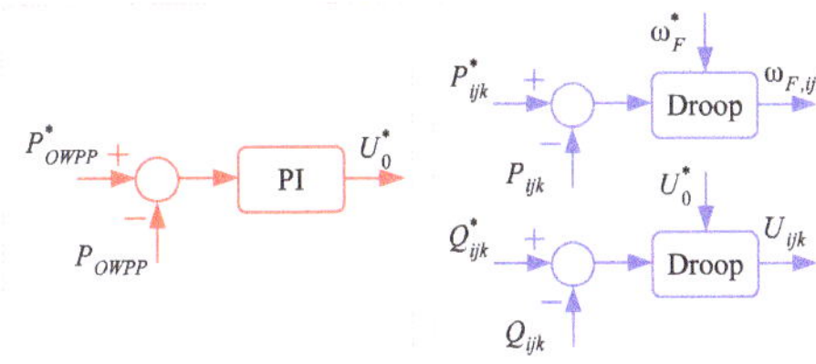

Figure 4. Left: centralized power control; right: distributed droop controls.

The centralized controller was based on a PI controller operating at a 20-ms sampling time. Using the total optimum power reference from the WTGs (P^*_{OWPP}) and the actual power being generated by the WPP (P_{OWPP}), the controller calculates the reference voltage (U^*_0) to be used by each individual WTG droop controller. ω^*_F is the grid frequency reference. A communication delay of 40 ms was considered for the WPP active power controller. The WPP active power controller had a bandwidth of 4 Hz.

Each WTG implemented a distributed P/ω and Q/V droop controller. The droop controller calculates the frequency ω_{Fijk} and the voltage U_{ijk} of the WTGs grid side converters (GSC) using the measured and reference active power P_{ijk} and the reactive power Q_{ijk}, respectively. Local P_{ijk} and Q_{ijk} measurements were filtered with a 10-rad/s low pass filter. More advanced controllers can also be implemented for P and Q sharing amongst wind turbines [23].

The AC-cable parameters, as well as all those of the system in Figure 3 are listed in the Appendix A.

5. Wind Power Plant Aggregation

This section includes the application of the proposed technique to the described case and its comparison with two well-established aggregation strategies.

5.1. Aggregation Based on Multi-Objective Optimization

The state space admittance model of each one of the wind power plants was obtained following the procedure explained in Section 2, including input reduction with σ equal to a vector of ones. The state space model thus obtained was of order 150.

Before carrying out the multi-objective optimization, it is important to ensure that the full system can be reduced to a system with only three states (according to Equations (10) and (11)). To this avail, the contribution of each state has been analyzed obtaining the Hankel singular values of the system. Figure 5 shows that the dynamics of the complete system are mainly dominated by three states; therefore, a third order reduced system can effectively capture most of the dynamic behavior of the full-order system.

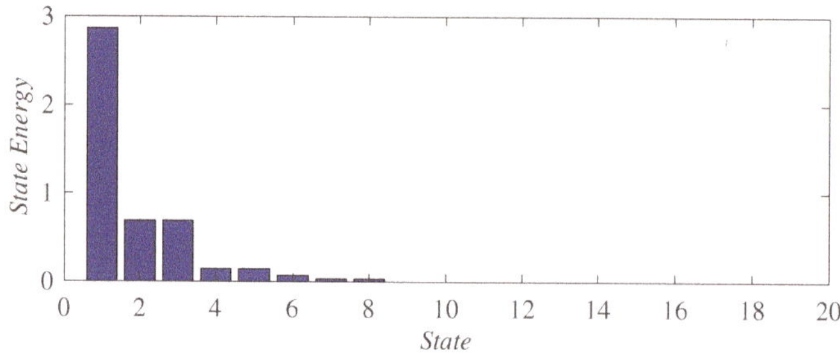

Figure 5. Hankel singular values (state contributions; the first 20 states are shown).

To carry out the multi-objective optimization problem, the weights W_i and the restrictions on R_{Teq}, L_{Leq}, and R_{Leq} need to be defined. To set the weights W_i properly, it is important to notice that the first objective used in the optimization function (12) is related to a frequency, while the other two are related to gains. Therefore, the weight W_i related to the first optimization objective is 1000-times higher than the other two weights. The selected weights W_i are shown in the Appendix A.

The parameters R_{Teq}, L_{Leq}, and R_{Leq} were restricted to be always positive. Moreover, their limits can be further refined. Restriction on the equivalent resistance can be obtained from the objective $K^*_{\omega=0}$. Analyzing the circuit of Figure 2, we know that the sum of $R_{Teq} + R_{Leq}$ shall be equal to $1/K^*_{\omega=0}$. Therefore, the maximum value of the addition of both resistors was $1/K^*_{\omega=0}$.

The restrictions on L_{Leq} were calculated by considering the simplified expression for the cable's first resonant frequency:

$$\omega_{res} = \sqrt{\frac{1}{LC}} \rightarrow L_{ini} = \frac{1}{C_{eq}\omega_{res}^2} \tag{14}$$

where C_{eq} is the total array cable capacitance and ω_{res} is the first cable resonance frequency obtained from the detailed state space admittance model. Therefore, the restrictions for L_{Leq} were set to $[0.01L_{ini}, 100L_{ini}]$. The selected parameter ranges are shown in the Appendix A.

With all the restrictions set, the multi-objective optimization problem was solved, and the obtained parameters are shown in Table 1. Figure 6 shows the admittance of the obtained aggregated model as a function of frequency.

Table 1. Equivalent parameters obtained from different aggregation methods.

Parameters	Voltage Drop	Power Losses	System Reduction
L_T (mH)	3.0143	3.0143	2.825483
R_T (Ω)	0.094696	0.094696	0.088765
C_L (μF)	18.135	18.135	18.135
L_L (mH)	0.41342	0.31412	0.509834
R_L (Ω)	0.08384	0.057677	0.059011

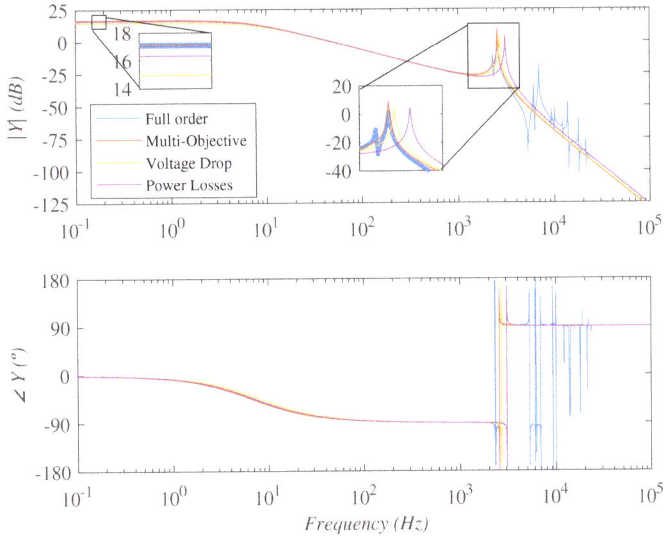

Figure 6. WPP grid admittance as a function of frequency: blue: 150 order system; red: multi-objective optimization; yellow: voltage drop; purple: power loses.

5.2. Aggregation Based on Voltage Drop

The voltage drop aggregation method presented in [12] was proposed in order to simplify frequency response analysis for WTG interaction with the grid. The following equations are a summary of the aggregation method in [12]. When WTGs are connected in a string, the total equivalent string impedance is:

$$Z_{ik} = \frac{1}{m_i} \sum_{j=1}^{m_i} j Z_{ijk} \tag{15}$$

where Z_{ik} is the equivalent R-L impedance corresponding to the ith string of the kth WPP and Z_{ijk} is the cable R-L impedance between WTGs j and $j+1$. Equation (16) was used for the parallel connection of several strings.

$$Z_k = \frac{1}{\sum_{i=1}^{n} \frac{1}{Z_{ik}}} \tag{16}$$

where Z_k is the equivalent impedance of WPP k. The aggregated capacitance of the system was obtained as follows:

$$C_k = \sum_{j=1}^{m_i} \sum_{i=1}^{n} C_{ijk} \tag{17}$$

where C_{ijk} is the cable capacitance between WTGs j and $j+1$. Additionally, the WTG transformer impedance and PWM filter were scaled by using their per unit values and taking into consideration how many WTGs were in operation.

Following these steps, an aggregated model of the system was calculated. The obtained values are shown in Table 1.

5.3. Aggregation Based on Power Losses

WTG aggregation based on power losses was used for load-flow studies and also in electromechanical (RMS) stability. Moreover, this aggregation technique has been applied successfully in many real-life wind farm projects [7].

The power loss aggregation technique proposed in [9] was based on the following equations:

$$Z_{ik} = \frac{1}{m_i^2} \sum_{j=1}^{m_i} j^2 Z_{ijk} \tag{18}$$

$$Z_k = \frac{\sum_{i=1}^{n} m_i^2 Z_{ik}}{[\sum_{i=1}^{n} m_i]^2} \tag{19}$$

$$C_k = \sum_{j=1}^{m_i} \sum_{i=1}^{n} C_{ijk} \tag{20}$$

where Z_{ik} is the equivalent R-L impedance corresponding to the ith string of the kth WPP, Z_{ijk} is the R-L cable impedance between WTGs j and $j + 1$, C_{ijk} is the cable capacitance between WTGs j and $j + 1$, and C_k is the total capacitance of WPP k.

In the same way as the voltage drop-based aggregation, WTG transformer impedance and PWM filter were scaled from their pu values, taking into consideration how many turbines were in operation.

The values obtained for this aggregation method applied to the considered case study are shown in Table 1.

5.4. Analytical Frequency Response of Different Aggregation Techniques

Table 1 shows the resulting parameters of the aggregated models obtained with the proposed aggregation technique and with two alternative state-of-the-art techniques.

Figure 6 shows the WPP grid admittance as a function of the frequency of the complete 150-state SISO model, as well as that of each one of the aggregated models (multi-objective optimization, voltage drop, and power losses' aggregation). All aggregation techniques showed a relative good agreement with the detailed model admittance for a wide range of frequencies.

However, the proposed technique showed a much closer match of low frequency admittance values and, more importantly, an excellent match of both main resonant peak frequency and amplitude. It is worth noting that both voltage drop- and power loss-based aggregation showed main resonant peak frequencies that were far off from the actual resonant peak. The largest error was obtained from the power loss error, which led to a resonant peak frequency of more than 500 Hz higher than its actual value.

Additionally, different WPP topologies have been studied, in order to verify that the proposed aggregation technique is valid for different WPP configurations. The studies covered different numbers of connected strings and also different numbers of connected WTGs per string. In all cases, the multi-objective aggregation technique always showed a better admittance frequency response match than the other two considered aggregation techniques.

Clearly, the results in Figure 6 represent small signal operation; however, the proposed aggregation technique had a very good agreement with the actual grid admittance for a very wide range of frequencies, including the main resonant peak. Therefore, the proposed aggregation technique is expected to show dynamic characteristics very close to the detailed 150-state system. The dynamic response of the proposed and state-of-the-art techniques is covered in Section 6.

6. Results

The frequency results in Figure 6 suggest that the proposed aggregation method should be able to represent the full system dynamics with less error than the other two considered aggregation techniques. Therefore, the EMTsimulation of the considered test cases was carried out in order to verify the dynamic behavior of each aggregation method.

Four scenarios have been considered, one with full detail models of each cluster (considering the full 150 WTGs of 8 MW each), and considering an aggregated equivalent for each one of the three

400-MW clusters (with multi-objective optimization, voltage drop, and power losses' aggregation methods).

The PSCAD-EMT simulations for the system shown in Figure 3 have been carried out in all four scenarios considered. However, for the sake of clarity, only the full simulation results corresponding to the detailed and multi-objective aggregated model are shown in the simulation section. The results corresponding to the other two aggregation methods are shown as errors with respect to the detailed model simulations.

The validation of the aggregated models has been carried out by comparing the active and reactive power and voltage and current (P_{Fk}, Q_{Fk}, V_{Fk}, I_{Fk}) responses at the points of common coupling of each one of the wind power plants (buses $PCC_{1,2,3}$ in Figure 3).

Three test cases have been considered:

- Fast changes in active power reference.
- Disconnection of one of the three wind power plants.
- Disconnection and re-connection of diode rectifier station AC filters.

Finally, a study of the computational load of each alternative has been carried out.

6.1. Simulation Results

6.1.1. Case 1: Fast Changes in Active Power Reference

The three wind power plants were initially operated in islanding mode (DR not conducting) with a voltage reference of 0.87 pu. From this state, the active power reference was ramped up from 0–1 pu in 150 ms for all three WPPs at the same time. Once all the WPPs reached their steady state operation, several step transients on active power reference were applied from 1 down to 0.5 pu, by sequentially reducing the active power reference P_{Fk}^* of each individual wind power plant.

Figure 7 shows the behavior of the active power P_{Fk}. Initially, the active power reference P_{Fk}^* rose to 1 pu at $t = 0.2$ s, but power production did not increase until currents began to flow through the DR at $t = 0.25$ s. The reason for this is that the AC-side DR voltage had to be higher than 0.9 pu for the DR station to start conducting. Then, at $t = 0.8$ s, the active power reference P_{Fk}^* was decreased to 0.5 pu in each cluster every 0.7 s. Such rapid active power transients might not be realistic, but have been chosen as an extreme case for validation purposes. Figure 7 clearly shows that both aggregated and detailed models led to very similar active power dynamic results.

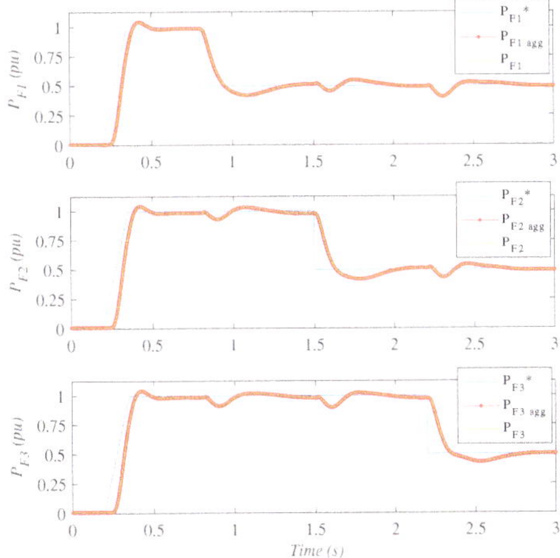

Figure 7. Case 1—Active power P_{Fk} at PCC-k of each diode rectifier platform for detailed and aggregated models.

Figure 8 shows the reactive power responses for each WPP Q_{Fk}. The reactive power produced by the AC filters was compensated by the WTGs while DRs were not delivering power. Conversely, when the DRs started conducting, the capacitor and AC filter banks' reactive power compensated that absorbed by the DRs, and hence, the reactive power Q_{Fk} delivered by the wind farms was reduced to a value very close to zero. Reactive power (Q_{Fk}) responses were also very similar for both aggregated and detailed models.

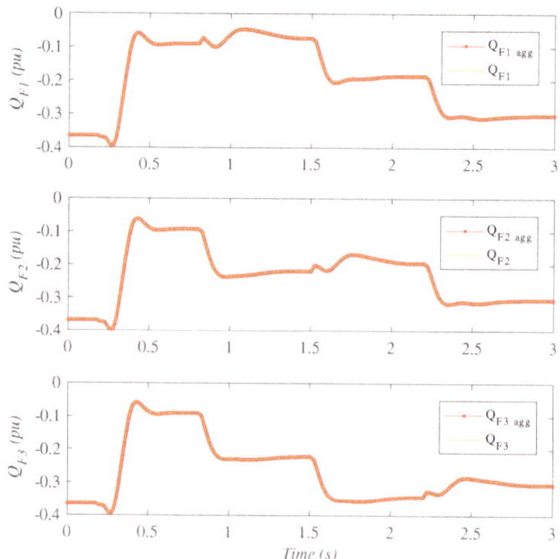

Figure 8. Case 1—Reactive power Q_{Fk} at PCC-k of each diode rectifier platform for the detailed and aggregated models.

Figure 9 shows the behavior of the voltage V_{Fk} at the PCC of each WPP. It shows clearly how active power flow by DRs depends of the voltage V_{Fk}. Figure 10 shows the magnitudes of current I_{Fk} at the PCC of each WPP. The current behavior was similar to that of the active power P_{Fk}. Figures 9 and 10 clearly show that the voltage and current dynamics from the aggregated and detailed models agree to a great extent.

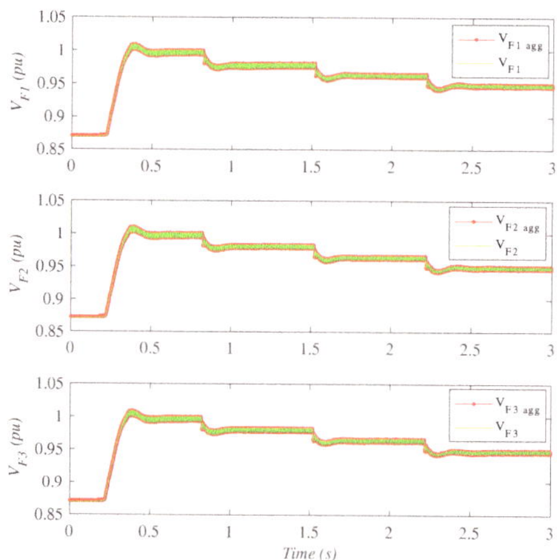

Figure 9. Case 1—Voltage amplitude at the PCC of each WPP (V_{Fk}) for detailed and aggregated models.

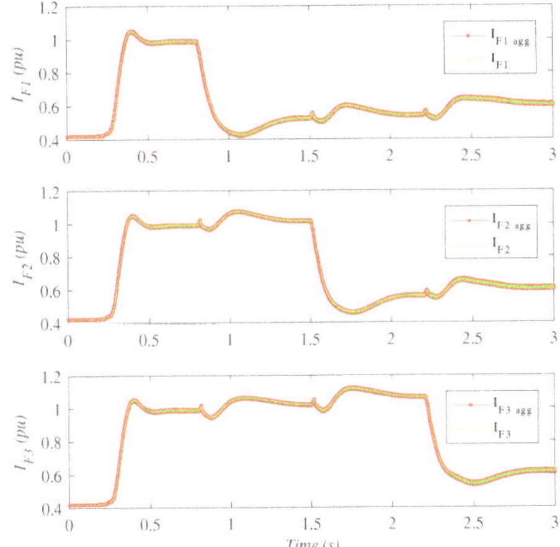

Figure 10. Case 1—Current amplitude at the PCC of each WPP (I_{Fk}) for detailed and aggregated models.

Three aggregation methods have been used to simulate the fast active power transients considered in Case 1. The results are shown in Figure 11, which shows the active and reactive power simulation errors between aggregated and detailed models at PCC-1, i.e., ($P_{F1Detailed} - P_{F1Agg}$) and ($Q_{F1Detailed} - Q_{F1Agg}$). Only the traces corresponding to PCC-1 are shown, as the other PCCs showed a very similar behavior.

The simulation error obtained using the proposed multi-objective optimization aggregation technique (in blue in Figure 11) was clearly lower than those with the alternative aggregation techniques.

The maximum active power simulation error with the proposed technique was about 0.3%, whereas the maximum simulation error for the reactive power simulation was approximately 0.2%. For the power loss technique, the active and reactive power maximum simulation errors were 0.9% and 0.5%, respectively, whereas for the voltage drop technique for the maximum simulation, the corresponding errors were 1.2% and 0.7%. Therefore, the proposed technique clearly showed a more accurate simulation during transients.

Moreover, for the complete case, the error variance using the multi-objective optimization aggregation technique was 0.0423, whereas the error variance with the voltage droop and power losses techniques was 0.4094 and 0.3120, respectively. Therefore, the error variance for the proposed technique was up to 9.5-times better than the other aggregation techniques.

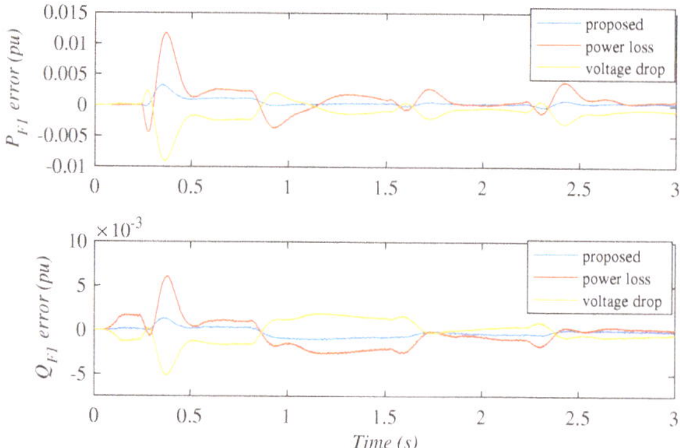

Figure 11. Case 1—Simulation errors for different aggregation techniques: top: $P_{F1Detailed} - P_{F1Agg}$; bottom: $Q_{F1Detailed} - Q_{F1Agg}$.

6.1.2. Case 2: Wind Power Plant Disconnection

The second test case consisted of the disconnection of WPP3, by opening the breaker at V_{F3} in Figure 3 at t = 0.05 s. Initially, all WPPs were generating 0.5 pu rated power. Figures 12 and 13 show the behavior of the WPP voltage and current magnitude at the PCC of each wind power plant.

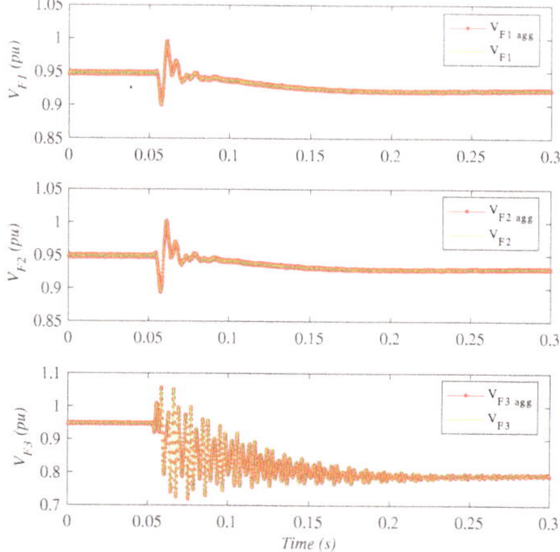

Figure 12. Case 2—Comparison of detailed and proposed aggregated simulations. WPP voltage V_{Fk}.

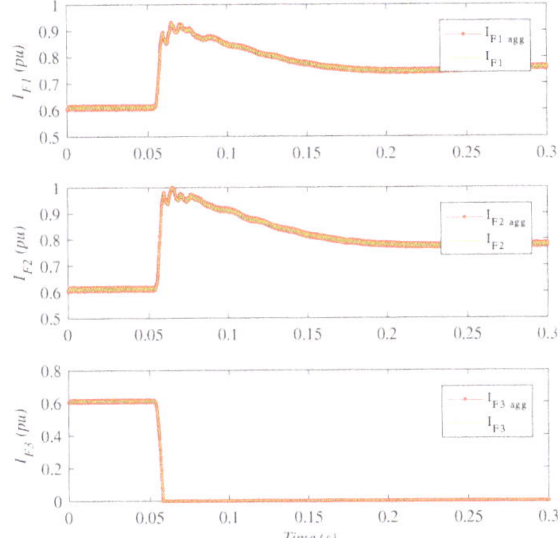

Figure 13. Case 2—Comparison of detailed and proposed aggregated simulations. WPP current I_{Fk}.

After disconnection, WPP3's voltage remained at its 0.8 pu reference value, as all considered WTGs were grid-forming (Figure 12). After the transient, the voltage of the WPPs that remained connected settled to a slightly smaller voltage, as the power transmitted through the diode rectifiers was now reduced by one third [5].

From Figures 12 and 13, it is clear that voltage and currents obtained from the detailed and from the proposed aggregated models showed an excellent agreement during the transient.

The same WPP3 disconnection case shown in Figures 12 and 13 has been repeated considering two alternative aggregation techniques (power loss and voltage drop). The results are shown in Figures 14 and 15, which show the difference between the detailed simulation and each one of the three considered aggregation techniques. Simulation errors are expressed as per unit with respect to rated values.

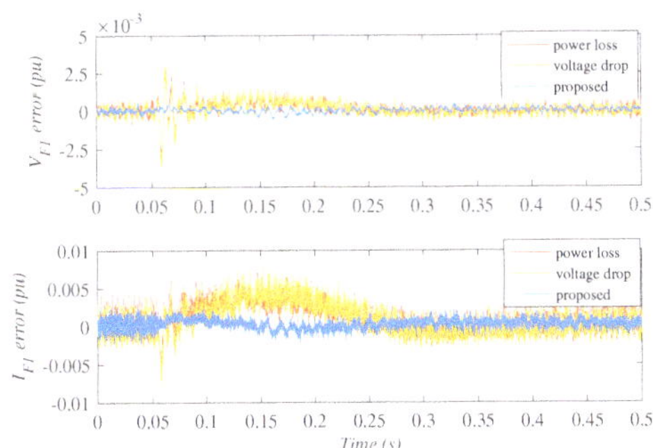

Figure 14. Case 2—Simulation errors for different aggregation techniques: top: $V_{F1Detailed} - V_{F1Agg}$; bottom: $I_{F1Detailed} - I_{F1Agg}$.

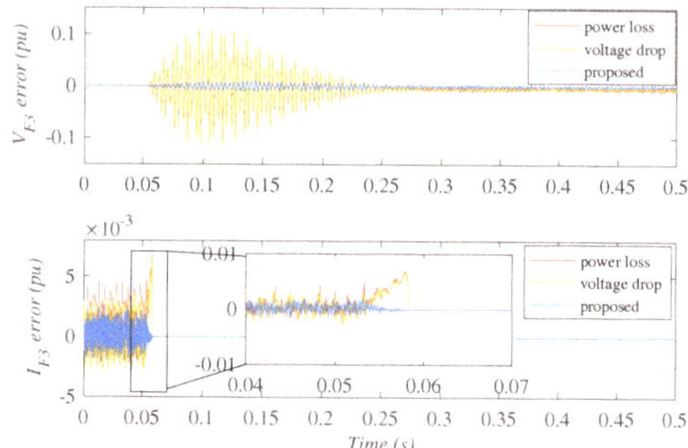

Figure 15. Case 2—Simulation errors for different aggregation techniques: top: $V_{F3Detailed} - V_{F3Agg}$; bottom: $I_{F3Detailed} - I_{F3Agg}$.

Figure 14 shows the errors on the simulated voltages and currents for the detailed and aggregated models, $(V_{F1Detailed} - V_{F1Agg})$ and $(I_{F1Detailed} - I_{F1Agg})$ respectively. These voltages and currents correspond to one of the two WPP that was not disconnected (WPP1). The proposed aggregation technique (in blue) clearly showed a better performance than existing aggregation techniques, although all aggregation techniques showed relatively small V_{F1} and I_{F1} simulation errors.

However, the voltage and current transients were much larger in the disconnected WPP3, as previously shown in Figures 12 and 13. The top graph of Figure 15 shows that the V_{F3} simulation error for the proposed aggregation method was much smaller than that of the existing aggregation methods (which even reached simulation errors larger than 10%).

On the other hand, the I_{F3} simulation errors were very small for all aggregation techniques, with only a small deviation just at the instant of WPP3 disconnection (Figure 15 bottom graph).

6.1.3. Case 3: Disconnection and Re-Connection of Diode Rectifier AC Filters

The third case consisted of the disconnection and re-connection of the AC capacitor and filter banks of DRU_3 (Figure 3). The capacitor and filter banks of each DRU were rated at 0.4 pu (160 MVAr) and were connected in several steps. However, this transient considered the connection and disconnection of the full AC filter banks, in order to validate the dynamic performance of the aggregated models.

The voltages and current output of each WPP during this test case are shown in Figures 16 and 17. The DRU_3 AC filter banks were disconnected at t = 0.1 s and re-connected at t = 0.8 s. Clearly, the harmonic contents of both voltage and current increased when the filter banks were disconnected. Moreover, voltage and current distortion was larger for WPP3, as it was electrically closer to the DRU station with no AC filters connected. WPP1 showed a relatively small voltage and current ripple when the filters were disconnected, whereas WPP2 showed intermediate harmonic contents.

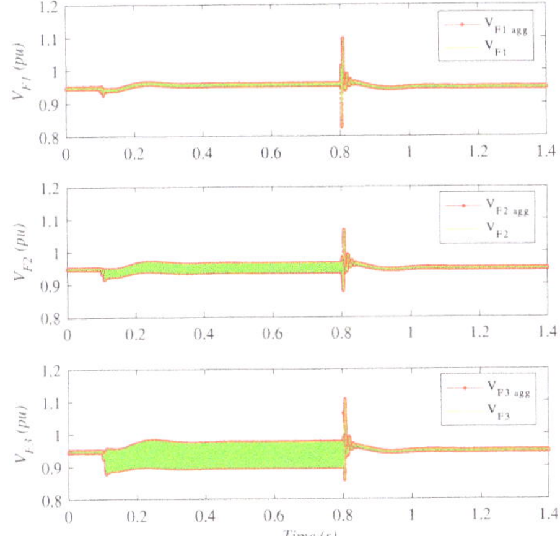

Figure 16. Case 3—Comparison of detailed and proposed aggregated simulations. WPP voltage V_{Fk}.

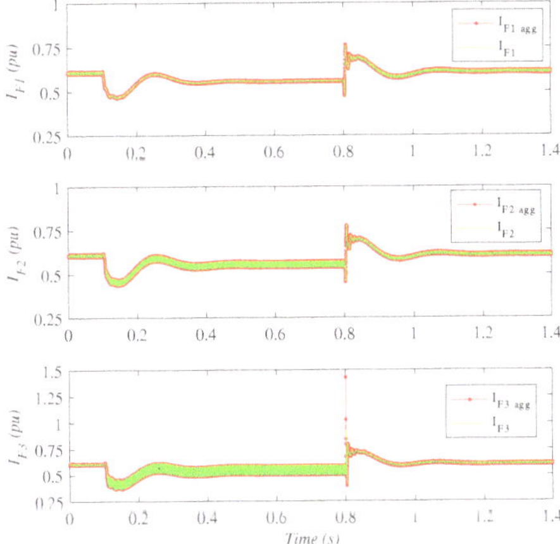

Figure 17. Case 3—Comparison of detailed and proposed aggregated simulations. WPP voltage I_{Fk}.

In any case, there was a very large agreement between detailed and aggregated simulations during the complete transient, except for the large current peak during AC filter bank reconnection in I_{F3} at t = 0.8 s (Figure 17), where the aggregated model overestimated the I_{F3} current peak.

The comparison of the proposed aggregation technique with existing techniques is shown in Figures 18 and 19. These figures show the simulation errors for the three aggregation techniques, during the DRU$_3$ AC filter disconnection and re-connection transient considered in this section.

Figure 18 shows the error on WPP3 active and reactive powers (P_{F3} and Q_{F3}) when simulated with each one of the considered aggregation techniques. All aggregation models adequately simulated the steady state active and reactive power delivered by WPP3. However, it was clear that, during the transients, the proposed aggregation technique (shown in blue) showed a smaller simulation error than existing techniques.

Figure 19 (top graph) shows that the proposed aggregation technique also showed a smaller V_{F3} error than other techniques when the filters were disconnected (i.e., between t = 0.1 and t = 0.8 s). Clearly, the proposed technique was relatively better at simulating V_{F3} voltage harmonics.

On the other hand, Figure 19 (bottom graph) shows that current I_{F3} simulation errors between t = 0.1 and t = 0.8 s were very similar for the three considered techniques, around 0.5% (albeit that the proposed technique error was slightly better).

It is worth noting that all aggregation techniques showed relatively large simulation errors for V_{F3} and I_{F3} when the AC filter banks were re-connected at t = 0.8 s (Figure 19). However, the proposed technique performed clearly better than voltage drop and power loss aggregation techniques.

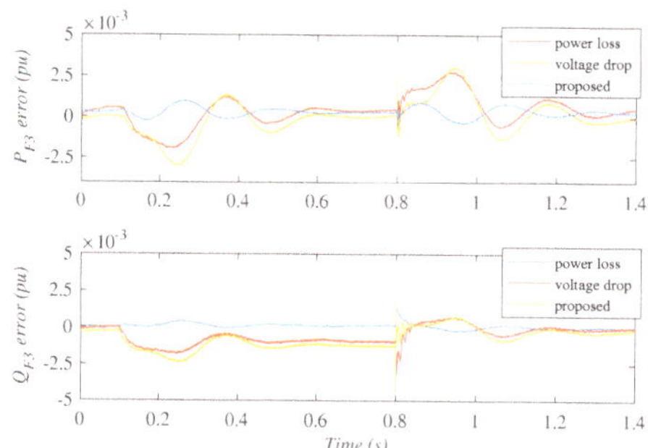

Figure 18. Case 3—Simulation errors for different aggregation techniques: top: $P_{F3Detailed} - P_{F3Agg}$; bottom: $Q_{F3Detailed} - Q_{F3Agg}$.

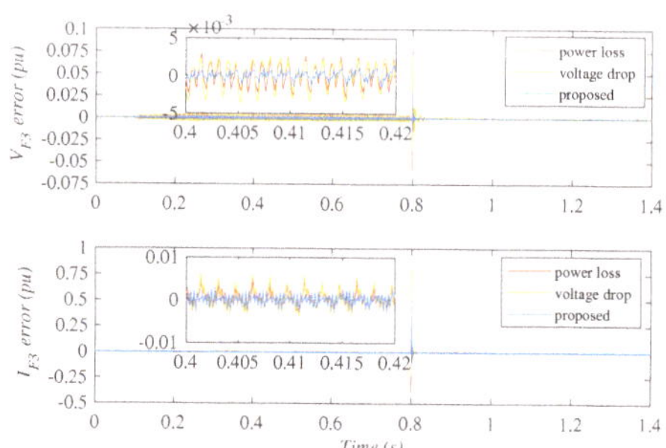

Figure 19. Case 3—Simulation errors for different aggregation techniques: top: $V_{F3Detailed} - V_{F3Agg}$; bottom: $I_{F3Detailed} - I_{F3Agg}$.

6.2. Simulation Performance

The simulations of all considered cases have been carried out with a 3.3-GHz Intel Core i7 PC with 16 GB DDR3 memory, SSD hard drive, and MS-Windows 7.

Considering a simulation run time of 3 s, Table 2 shows the complexity and simulation times of both detailed and aggregated models. The detailed simulation took 2 h, whereas the aggregated simulation was more than 350-times faster.

Therefore, the use of the proposed aggregated models for large Type-4 wind power plants was validated considering the similarity of the obtained results and the improvement on simulation times.

Table 2. Simulation times.

	WTGs	Nodes	Simulation Time
Detailed	150	2450	117 min
Aggregated	3	174	20 s

7. Discussion and Conclusions

This paper has presented an aggregation approach based on model order reduction using a multi-objective optimization technique. The proposed method consisted of first obtaining full admittance model of the WPP (in our case using state space techniques), then the inputs to the system were reduced to a single input applying the superposition principle, and finally, multi-objective optimization has been used to reduce the order of the system. The proposed aggregation technique considered voltage source WTGs' line side converters; therefore, it is also applicable to current-controlled voltage source converters.

Using the multi-objective optimization aggregation technique, the achieved aggregated admittance model had the same DC gain, main resonant frequency, and the same gain at the operating frequency as the detailed admittance model.

The proposed aggregation technique was validated considering the PSCAD/EMTDC simulation of a 1.2-GW HVDC DR-connected system, consisting of three 400-MW WPP of 50 Type-4 grid-forming WTGs each. Hence, the detailed system consisted of 150 individual wind turbines and 2450 nodes.

The proposed aggregation technique and two commonly-used aggregation strategies have been used to reduce the WPPs to only three aggregated WPPs (each one equivalent to 400 MW).

The comparison has been carried out considering three different test cases, namely fast active power reference changes, disconnection of a wind power plant, and disconnection and re-connection of DRU AC filter banks. These cases cover a wide range of dynamic and transient conditions.

The response against active power changes achieved by using the proposed aggregation method showed a worst case error three-times and an error variance 9.5-times better than aggregation techniques based on voltage drop or on power losses.

During WPP disconnection, all aggregation techniques showed relatively good PCC voltage and current simulation accuracy for the two wind power plants that remained connected. However, for the disconnected WPP, the proposed method showed voltage simulation accuracy 10-times better than standard methods, during the disconnection transient.

During disconnection and re-connection of one of the DRU AC filter banks, the proposed technique also showed better simulation accuracy regarding PCC voltage, active, and reactive power.

Finally, it has been shown that the aggregated simulations were 350-times faster than their detailed counterpart.

Therefore, when only the behavior at the points of common coupling is of interest, the proposed aggregated model provides important simulation time savings while delivering accurate results and preserving the main resonant characteristics of the full array system.

Author Contributions: The present work was developed with the following contributions: conceptualization, methodology, software, validation, formal analysis, research, writing, original draft preparation, and data curation: J.M.-T., S.A.-V., S.B.-P., and R.B.-G.; writing, review and editing, and supervision: S.A.-V., S.B.-P., and R.B.-G.

Funding: The authors would like to thank the support of the Spanish Ministry of Science, Innovation and Universities and EU FEDER funds under Grants DPI2014-53245-R and DPI2017-84503-R. This project has received funding from the European Union's Horizon 2020 research and innovation program under Grant Agreement No. 691714.

Acknowledgments: The authors would like to thank the editorial board, as well as the anonymous reviewers for their valuable comments that have greatly improved the quality of this document.

Conflicts of Interest: The authors declare no conflict of interest. The funders had no role in the design of the study; in the collection, analyses, or interpretation of data; in the writing of the manuscript; nor in the decision to publish the results.

Appendix A

System Parameters		
Wind Turbines		
Grid-side VSC: 1.2 kVcc, 690 Vac, 50 Hz PWM filter: $R_W = 0.008$ pu, $L_W = 0.18$ pu, $C_W = 20$ pu		
Transformer T_R: 8 MVA, 50 Hz, 0.69/66 kV (L-L rms), $X_T = 0.18$ pu, $R_T = 0.01$ pu		
Off-shore AC grid		
Base voltage V_F: 66 kV, 50 Hz		
Distance between WTs: 1.5 km		
Distance from cluster to DR platform: 3 km		
Cable section: a = 150 mm^2, b = 185 mm^2, c = 400 mm^2,		
String with 8 WT: a-a-a-b-b-b-c-c		
String with 7 WT: a-a-a-b-b-b-c		
DR stations		
Base voltage of rectifier transformer: 66/43 kV		
Filter and reactive power compensation bank for the 12-pulse rectifier according to CIGREbenchmark [24].		
HVDC system		
Base voltage of HVDC system: ±320 kV, 150 km		

Controllers		
PI power controller:	$K_P = 0.00006$	$T_I = 380$
Proportional frequency droop:	$m_P = 0.00125$ Hz/MW	
Proportional amplitude droop:	$n_P = 0.01$ V/MVAr	

Multi-objective optimization parameters	
Optimization objectives f_i^*:	[16,015.38 rad/s, 7.17725, 0.85227]
Weights W_i:	[1000.0, 1.0, 1.0]
R_{Teq} range:	[0.001, 0.139329] Ω
R_{Leq} range:	[0.001, 0.139329] Ω
L_{Leq} range:	[0.00215, 21.5] mH

References

1. Menke, P. *New Grid Access Solutions for Offshore Wind Farms*; EWEA Off-Shore: Copenhagen, Denmark, 2015.
2. Blasco-Gimenez, R.; Añó-Villalba, S.; Rodríguez, J.; Morant, F.; Bernal, S. Uncontrolled rectifiers for HVDC connection of large off-shore wind farms. In Proceedings of the 13th European Conference on. Power Electronics and Applications, Barcelona, Spain, 8–10 September 2009; pp. 1–8.
3. Blasco-Gimenez, R.; Añó-Villalba, S.; Rodríguez-D'Derlée, J.; Morant, F.; Bernal-Perez, S. Distributed Voltage and Frequency Control of Offshore Wind Farms Connected With a Diode-Based HVDC Link. *IEEE Trans. Power Electron.* **2010**, *25*, 3095–3105. [CrossRef]
4. Blasco-Gimenez, R.; Añó-Villalba, S.; Rodriguez-D'Derlée, J.; Bernal-Perez, S.; Morant, F. Diode-Based HVDC Link for the Connection of Large Offshore Wind Farms. *IEEE Trans. Energy Convers.* **2011**, *26*, 615–626. [CrossRef]
5. Bernal-Perez, S.; Añó-Villalba, S.; Blasco-Gimenez, R.; Rodriguez-D'Derlee, J. Efficiency and Fault Ride-Through Performance of a Diode-Rectifier- and VSC-Inverter-Based HVDC Link for Offshore Wind Farms. *IEEE Trans. Ind. Electron.* **2013**, *60*, 2401–2409. [CrossRef]

6. Blasco-Gimenez, R.; Aparicio, N.; Añó-Villalba, S.; Bernal-Perez, S. LCC-HVDC Connection of Offshore Wind Farms With Reduced Filter Banks. *IEEE Trans. Ind. Electron.* **2013**, *60*, 2372–2380. [CrossRef]
7. Kocewiak, L.H. Harmonics in Large Offshore Wind Farms. Ph.D. Thesis, Department of Energy Technology, Aalborg University, Aalborg, Denmark, 2012.
8. Kocewiak, L.H.; Hjerrild, J.; Bak, C.L. Wind turbine converter control interaction with complex wind farm systems. *IET Renew. Power Gener.* **2013**, *7*, 380–389. [CrossRef]
9. Muljadi, E.; Butterfield, C.P.; Ellis, A.; Mechenbier, J.; Hochheimer, J.; Young, R.; Miller, N.; Delmerico, R.; Zavadil, R.; Smith, J.C. Equivalencing the collector system of a large wind power plant. In Proceedings of the 2006 IEEE Power Engineering Society General Meeting, Montreal, QC, Canada, 18–22 June 2006.
10. Garcia-Gracia, M.; Comech, M.P.; Sallan, J.; Llombart, A. Modeling wind farms for grid disturbance studies. *Renew. Energy* **2008**, *33*, 2109–2121. [CrossRef]
11. Fernandez, L.M.; Garcia, C.A.; Saenz, J.R.; Jurado, F. Equivalent models of wind farms by using aggregated wind turbines and equivalent winds. *Energy Convers. Manag.* **2009**, *50*, 691–704. [CrossRef]
12. Brogan, P. The stability of multiple, high power, active front end voltage sourced converters when connected to wind farm collector systems. In Proceedings of the EPE Wind Energy Chapter Seminar, Brno, Czech Republic, 4–6 May 2010; pp. 1–6.
13. Kocewiak, L.H.; Hjerrild, J.; Bak, C.L. Wind farm structures' impact on harmonic emission and grid interaction. In Proceedings of the The European Wind Energy Conference & Exhibition, Warsaw, Poland, 20–23 April 2010.
14. Martínez-Turégano, J.; Añó-Villalba, S.; Chaques-Herraiz, G.; Bernal-Perez, S.; Blasco-Gimenez, R. Model aggregation of large wind farms for dynamic studies. In Proceedings of the IECON 2017—43rd Annual Conference of the IEEE Industrial Electronics Society, Beijing, China, 29 October–1 November 2017; pp. 316–321.
15. Semlyen, A.; Deri, A. Time Domain Modeling of Frequency Dependent Three Phase Transmission Line Impedance. *IEEE Trans. Power Appar. Syst.* **1985**, *PAS-104*, 1549–1555. [CrossRef]
16. Garcia, N.; Acha, E. Transmission line model with frequency dependency and propagation effects: A model order reduction and state-space approach. In Proceedings of the 2008 IEEE Power and Energy Society General Meeting—Conversion and Delivery of Electrical Energy in the 21st Century, Pittsburgh, PA, USA, 20–24 July 2008; pp. 1–7.
17. Beerten, J.; D'Arco, S.; Suul, J.A. Frequency-dependent cable modelling for small-signal stability analysis of VSC-HVDC systems. *IET Gener. Transm. Distrib.* **2016**, *10*, 1370–1381. [CrossRef]
18. Ruiz, C.; Abad, G.; Zubiaga, M.; Madariaga, D.; Arza, J. Frequency-Dependent Pi Model of a Three-Core Submarine Cable for Time and Frequency Domain Analysis. *Energies* **2018**, *11*, 2778. [CrossRef]
19. Brayton, R.; Director, S.; Hachtel, G.; Vidigal, L. A new algorithm for statistical circuit design based on quasi-Newton methods and function splitting. *IEEE Trans. Circuits Syst.* **1979**, *26*, 784–794. [CrossRef]
20. Gembicki, F.W. Vector Optimization for Control with Performance and Parameter Sensitivity Indices. Ph.D. Thesis, Case Western Reserve University, Cleveland, OH, USA, 1974.
21. Fleming, P.J. Computer aided control system design using a multi-objective optimization approach. In Proceedings of the Control 1985 Conference, Cambridge, UK, 22 January 1985; pp. 174–179.
22. Han, S.P. A globally convergent method for nonlinear programming. *J. Optim. Theory Appl.* **1977**, *22*, 297–309. [CrossRef]
23. Guerrero, J.M.; Vasquez, J.C.; Matas, J.; de Vicuna, L.G.; Castilla, M. Hierarchical Control of Droop-Controlled AC and DC Microgrids—A General Approach Toward Standardization. *IEEE Trans. Ind. Electron.* **2011**, *58*, 158–172. [CrossRef]
24. Szechtman, M.; Wess, T.; Thio, C.V. First benchmark model for HVDC control studies. *Electra* **1991**, *135*, 54–73.

Article

Small Wind Turbine Emulator Based on Lambda-Cp Curves Obtained under Real Operating Conditions

Camilo I. Martínez-Márquez, Jackson D. Twizere-Bakunda, David Lundback-Mompó, Salvador Orts-Grau *, Francisco J. Gimeno-Sales and Salvador Seguí-Chilet

Instituto Interuniversitario de Investigación de Reconocimiento Molecular y Desarrollo Tecnológico, Universitat Politècnica de València, Camino de Vera 14, 46022 Valencia, Spain; camarma3@posgrado.upv.es (C.I.M.-M.); jtwizbakd@gmail.com (J.D.T.-B.); dalunmom@gmail.com (D.L.-M.); fjgimeno@eln.upv.es (F.J.G.-S.); ssegui@eln.upv.es (S.S.-C.)
* Correspondence: sorts@eln.upv.es; Tel.: +34-963-877-000

Received: 17 May 2019; Accepted: 23 June 2019; Published: 26 June 2019

Abstract: This paper proposes a new on-site technique for the experimental characterization of small wind systems by emulating the behavior of a wind tunnel facility. Due to the high cost and complexity of these facilities, many manufacturers of small wind systems do not have a well knowledge of the characteristic λ-C_p curve of their turbines. Therefore, power electronics converters connected to the wind generator are usually programmed with speed/power control curves that do not optimize the power generation. The characteristic λ-C_p curves obtained through the proposed method will help manufacturers to obtain optimized speed/power control curves. In addition, a low cost small wind emulator has been designed. Programmed with the experimental λ-C_p curve, it can validate, improve, and develop new control algorithms to maximize the energy generation. The emulator is completed with a new graphic user interface that monitors in real time both the value of the λ-C_p coordinate and the operating point on the 3D working surface generated with the characteristic λ-C_p curve obtained from the real small wind system. The proposed method has been applied to a small wind turbine commercial model. The experimental results demonstrate that the point of operation obtained with the emulator is always located on the 3D surface, at the same coordinates (rotor speed/wind speed/power) as the ones obtained experimentally, validating the designed emulator.

Keywords: wind turbine emulator; wind turbine energy systems; renewable energies

1. Introduction

Although wind is difficult to predict accurately because of the many factors on which it depends, its renewable nature and abundance make it a good candidate for providing energy on a large scale. In recent decades, the technology necessary to use wind energy for generating electricity has been developed. Aerodynamic study for the capture of wind power [1], design of the electric generators and power converters [2–6], control of the system, or the optimization of energy generation [6–10] are just a few examples of the challenges that still face engineers.

Testing wind energy conversion systems under real operating conditions requires a wind generator installed on a site with good wind conditions or a wind tunnel facility. Both options are unaffordable for manufacturers of small wind generators or academic researchers due to the magnitude of these installations and their cost. However, when no aerodynamic studies are being performed, e.g., design and testing of power electronic converters and electric generators, a more economically viable solution is to implement a wind turbine emulation system. In an emulator, the wind turbine is replaced by an electromechanical actuator that incorporates in the control board the model of the wind turbine to be emulated. The electromechanical actuator is mechanically coupled, on a bench, to the electrical generator under study. A wind turbine emulation system has several advantages: easy implementation;

reduced cost; less space needed; simple operation; and the possibility of setting any desired wind profile at the right time without dependence on weather forecasts.

Several implementations of wind emulators are detailed in References [11–21]. Wind emulators are classified mainly according to the type of electromechanical actuator used, or by the parameters considered for the implementation of the aerodynamic model. DC motors can be used as electromechanical actuators [11–15] because of their easy torque control through the armature current. However, DC motors present some important drawbacks such as: considerable size-power ratio, high cost, the need for maintenance, etc. [15]. Other electromechanical actuators use induction motors [16–19], permanent magnet synchronous motors (PMSM) [20], or servomotors [21]. The effect of the variation of wind speed with height, known as wind shear, is included into the aerodynamic model of the turbine in Reference [20]. Other phenomena, such as the tower shadow effect (which occurs every time one of the blades passes in front of the tower) are also integrated into the turbine model in References [19,20]. A servomotor controlled by an inverter is used in Reference [21]. The inverter control signal is calculated using the generated wind torque feedback signal and the desired wind speed in the aerodynamic model. A wind system connected to the grid is simulated in Reference [19], using an induction motor as actuator. The torque to be developed by the actuator is calculated based on the wind speed and considering the tower shadow effect. Wind profiles are obtained offline by means of an anemometer located on the roof of the laboratory.

One important drawback of wind turbine emulators is the instability that appears during the compensation of the moment of inertia of the mechanical parts [17]. To generate the mechanical torque as accurately as possible, both in steady state and during transients, the adjustment of the moment of inertia was obtained by compensation in Reference [13], where different transient responses of the emulator are tested, simulating variations in the wind speed and considering different values of the moment of inertia. However, the variation of the power coefficient $\left(C_p\right)$ during transients was not considered in the wind emulator. Another common disadvantage of these turbine emulation systems is that they constrain the performed emulation to theoretical mechanical models, making the overall system less realistic. Many manufacturers of small wind systems do not have good knowledge of the λ-C_p characteristic curve of their turbines, due to the high cost and complexity of a wind tunnel. Therefore, power electronics converters connected to these small wind generators are usually programmed with speed/power control curves that do not optimize the power generation.

This work proposes a new on-site technique for the experimental characterization of small wind systems by emulating the behavior of a wind tunnel facility. The λ-C_p characteristic curves obtained through the proposed method will help manufacturers to obtain optimized speed/power control curves. In addition, a low cost small wind emulator has been designed. Programmed with the experimental λ-C_p characteristic curve, it can validate, improve, and develop new control algorithms to maximize the energy generation of the small wind system.

The emulator is completed with a novel graphic user interface (GUI). This GUI allows the real-time visualization of the operating point of the turbine, both in the lambda—power coefficient (λ-C_p) curve and in the working surface of the turbine. The working surface of the turbine represents the possible operating points of the turbine and is obtained from the projection of the λ-C_p characteristic curve in a 3D space given by the following variables: available electric power (in W); wind turbine angular speed (in rpm); and wind speed (in m/s).

This article is organized as follows. Section 2 presents the modeling of the small wind turbine and the mechanical system. Section 3 describes the approach for the estimation of the λ-C_p curve based on the analysis of experimental data acquired in several small wind turbines operating under real outdoor conditions. Section 4 presents the proposed wind emulator that is combined with a novel graphic user interface which enables obtaining, among other things, the working surface of the turbine. This section also includes the results of the developed experiments that show the capabilities of the proposed small wind emulator. Finally, Section 5 summarizes the most important points of the proposals developed in this paper.

2. Wind Turbine and Mechanical System Modeling

Figure 1 depicts a generic wind generator system. The wind turbine should have two or three blades and can be directly coupled to a permanent magnet synchronous generator (PMSG) or by means of a gearbox. The incoming aerodynamic turbine torque (T_T), produced by the action of the wind on the blades, is finally applied to the electrical generator, which in turn will generate a counter-electromagnetic torque (T_G). The moment of inertia (J) is the sum of the moments of inertia of both, the turbine (J_T), and the electrical generator (J_G):

$$J = J_T + J_G \tag{1}$$

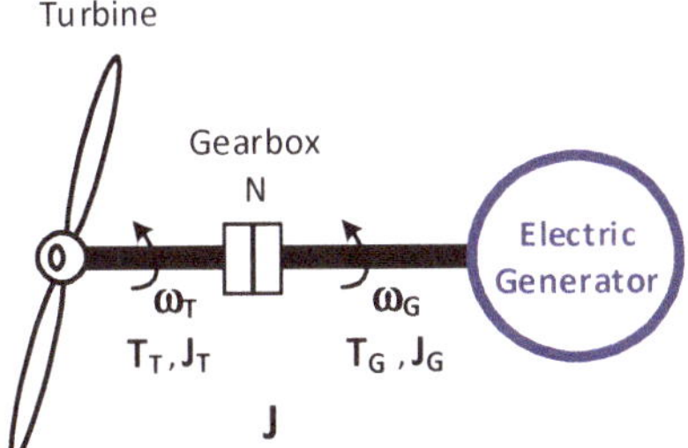

Figure 1. Mechanical model of a generic wind generation system.

Certain effects, like viscous friction coefficient and shaft torsional spring constant, are usually neglected to simplify the model. By means of this approximation, the mechanical model of the turbine can be expressed as follows:

$$\frac{T_T}{N} - T_G = J\frac{d\omega_G}{dt} \tag{2}$$

where ω_G is the angular speed of the electric PMSG generator and N is the gear ratio of the gearbox. The power captured by a wind turbine depends on the interaction between the wind and the turbine blades. The power available in the wind (P_{wind}) is expressed as follows [5]:

$$P_{wind} = \frac{1}{2}\rho A V_{wind}^3 \tag{3}$$

where ρ is the density of the air (1225 kg/m³), A is the area swept by the turbine blades given in m², and V_{wind} is the wind speed given in m/s. The power coefficient C_p is the ratio of the power captured by the wind turbine (P_T) on the low-speed shaft (P_{LSS}) to the power available in the wind, being expressed as follows:

$$C_p = \frac{P_{LSS}}{P_{wind}} = \frac{P_T}{P_{wind}} \tag{4}$$

Thus, the expression of P_T can be obtained by combining (3) and (4) as follows:

$$P_T = \frac{1}{2}C_p\rho A V_{wind}^3 \tag{5}$$

The maximum C_p achievable by a turbine under ideal conditions is 0.59 (Betz limit). The power coefficient C_p is a function of two parameters: the tip–speed ratio (λ) and the pitch angle (β) of the blades. The tip–speed ratio relates the blade tip speed with the wind speed, and is expressed as follows:

$$\lambda = \frac{V_T}{V_{wind}} = \frac{\omega_T \cdot r_T}{V_{wind}} \tag{6}$$

where V_T is the linear speed at the tip of the turbine blades, ω_T is the angular speed of the rotor of the turbine in rad/s, and r_T is the length of the blade in m. The aerodynamic characteristic of a wind turbine can be modified by varying the pitch angle (β) of the blades. Such a technique, denoted as pitch control, is usually applied in large wind turbines, but is uncommon in small wind turbines.

Figure 2 shows a set of λ-C_p curves for different values of β, obtained using a non-lineal function [22]. Each curve has a different maximum power coefficient value, $C_{p_{max}}$, at a different λ value (denoted as optimal lambda or λ_{opt}).

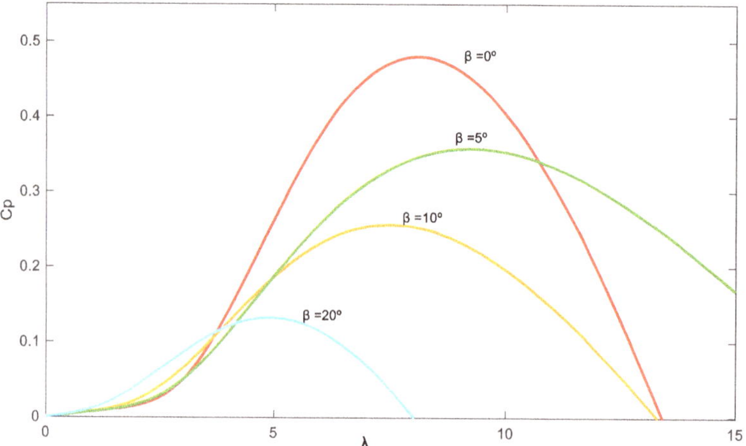

Figure 2. Variation of the λ-C_p curve depending on the pitch angle of the blades (β).

For small wind turbines, without pitch control, $\beta = 0$. From Equation (5), the aerodynamic torque developed by a small wind turbine can be expressed as follows:

$$T_T = \frac{1}{2\omega_T} C_p \rho A V_{wind}^{3} \tag{7}$$

The operating point of the wind turbine depends on the electrical power consumed by the electrical loads, as detailed in Equation (2).

3. Experimental Method for the Estimation of the λ-C_p Characteristic Curve of a Small Wind Turbine

Due to the high cost of the necessary tests, manufacturers do not offer the λ-C_p characteristic curve of their small wind turbines. A knowledge of λ-C_p characteristics is crucial for developing an efficient control of the energy conversion process, and securing the safe operation of the wind turbine. This section presents an experimental method developed for the estimation of the λ-C_p characteristic curve of a small wind turbine without pitch control ($\beta = 0$). This method is intended to give manufacturers of small wind turbines a practical and cheap tool to obtain the λ-C_p characteristic curve of their turbines.

Given the difficulties of measuring P_T in a commercial system and the low losses in the conversion from mechanical to electrical power when compared with the output electrical power, the value of C_p is usually calculated using the electrical generated power (P_e) [23]:

$$C_p = \frac{P_T}{P_{wind}} \cong \frac{P_e}{P_{wind}} \tag{8}$$

The λ-C_p characteristic curve will be obtained after processing the data logged from several small wind turbines operating under real outdoor conditions. All of them are equipped with a specific firmware control that performs the test conditions required to acquire the following magnitudes: wind speed (V_{wind}); turbine angular speed (ω_T); and electrical generated power (P_e).

The small wind turbine used in the experimental parts of the paper is a Bornay Wind Plus 25.3+. This turbine is equipped with a set of three blades of $r_T = 2$ m and has a moment of inertia $J = 5.75$ Kg·m^2. The value of J has been provided by the manufacturer and is the moment of inertia to be emulated. The turbine starts its operation for $V_{wind} > 3$ m/s, develops the rated power for $V_{wind} = 12$ m/s, and includes an automatic brake system if $V_{wind} > 14$ m/s. The wind turbine is directly coupled to the shaft of the electrical generator, so N = 1. The electrical generator used in the Bornay Wind Plus 25.3+ is a permanent magnet synchronous generator (PMSG) and has the rated values shown in Table 1.

Table 1. Technical characteristics of the electrical generator used in the small wind turbine.

Voltage	Rated Power (kW)	Peak Power (kW)	Rated Speed (rpm)	Pole Pairs
3–220 V_{L-L}	5 kW	7 kW	375	8

Data is collected from three different small wind turbines, as part of three different off-grid systems. The presented results correspond to a two-month period of data recording. For data collection purposes, the control system of the turbine is modified to perform a speed reference (ω_T^*) sweep in a loop, varying from $\omega_{T_{min}}^* = 90$ rpm to $\omega_{T_{max}}^* = 562.5$ rpm ($1.5\omega_{T\_rated}$) with a 10 rpm increment every 30 min. Meanwhile, the system is sampling, every second, the three magnitudes needed to obtain the λ-C_p points (V_{wind}, P_e, and ω_T). To maintain ω_T constant, the electronic converter connected in the AC output of the PMSG must be able to extract all the generated electrical power. A controlled resistive load is used to extract the excess of power that the loads are not demanding. The data collection algorithm considers safety issues, resetting the speed reference to $\omega_{T_{min}}^*$ if the incoming torque exceeds a generator torque threshold that is set to 90% of the rated torque during the data recording. This limitation ensures that the maximum electrical torque of the generator is not exceeded, and the system can still be controlled. For every ω_T^* value, the system collects a huge amount of data that has to be filtered to discard non-significant values. Figure 3 depicts the block diagram of the acquisition system and the flowchart of the proposed algorithm for the estimation of the λ-C_p characteristic curve.

A filtering process is started after the data recording. Since the wind turbines have a cut-in speed, all data collected corresponding to $V_{wind} < 3$ m/s is discarded. The following filtering process is to delete the data corresponding to the transients, e.g., during wind speed changes. A set of values of V_{wind}, P_e, and ω_T is considered valid if the turbine angular speed is equal to its reference ($\omega_T = \omega_T^*$) and if V_{wind} and ω_T are stable. The stability of V_{wind} and ω_T is important because the turbine can store kinetic energy during transients. V_{wind} and ω_T stability are verified by comparing the acquired sample (k) with the $k - 4$ and $k + 4$ samples (previous and subsequent recorded values, respectively). Maximum variation in V_{wind} during the comparison is limited to ± 1 m/s, while ω_T variation must be smaller than ± 5 rpm.

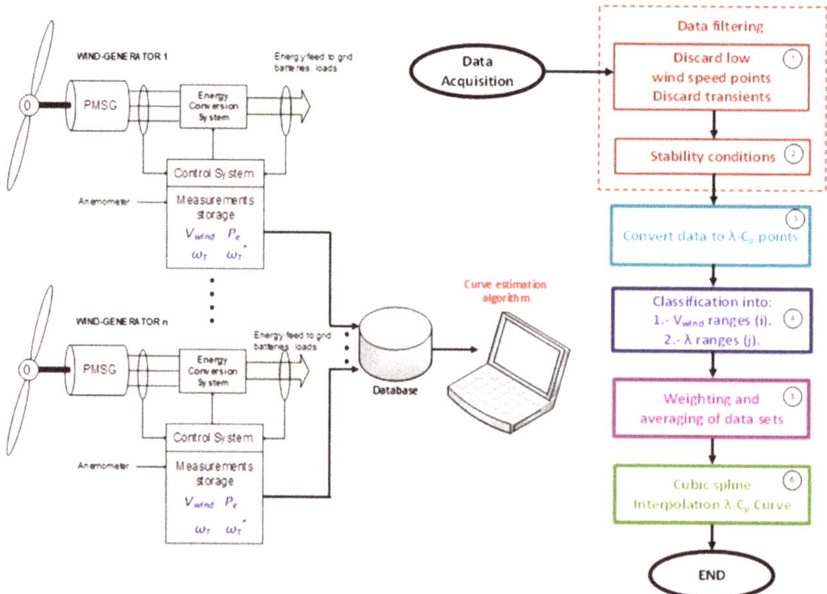

Figure 3. Block diagram of the data acquisition (**left**) and flowchart of the proposed method (**right**).

After the filtering process, the remaining sets of values of V_{wind}, P_e, and ω_T are processed to obtain the corresponding λ-C_p coordinates. It is important to notice that the altitude has been considered to correct for air density when C_p values are calculated. Figure 4 shows the λ-C_p coordinates obtained after the filtering process. The points corresponding to each turbine are shown with different colors and symbols.

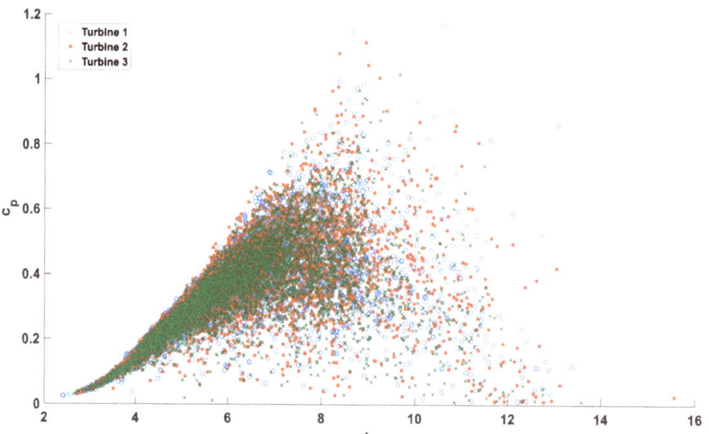

Figure 4. Raw experimental data obtained from the wind turbines.

The obtained λ-C_p points are classified firstly in V_{wind} ranges, being denoted by subscript m in Figure 5. These ranges are selected in this study with $\Delta V_{wind} = 0.5$ m/s, with acceptable values between 3 m/s (the cut-in speed of the turbine) to 16 m/s ($m = 1 \ldots 26$). The points corresponding to each V_{wind} range are classified again, this time in ranges of λ, being denoted by subscript n in Figure 5. These λ

ranges are selected to give enough resolution to the curve, $\Delta\lambda = 0.1$ being used in the experimental analysis, with a $\lambda_{max} = 16$ for the turbines and locations used in the study. The three-dimensional space defined by λ-C_p-V_{wind} is divided in m by n cuboids, as appears in Figure 5. A cuboid that contains less than five points is discarded because they are non-representative points of the λ-C_p curve.

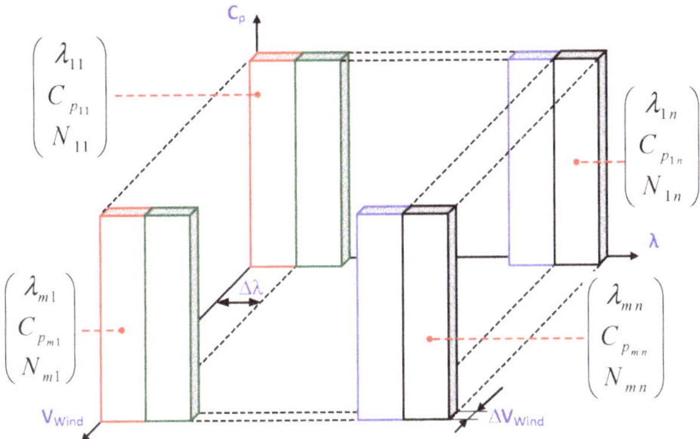

Figure 5. Description of the data classification in V_{wind} and λ ranges before the averaging and weighting process.

The total number of λ-C_p-V_{wind} points in each cuboid is an element N_{ij} of matrix **N**, where $i = 1..m$ and $j = 1..n$, expressed as follows:

$$\mathbf{N} = \begin{pmatrix} N_{11} & \cdots & N_{1n} \\ \vdots & \ddots & \vdots \\ N_{m1} & \cdots & N_{mn} \end{pmatrix} \tag{9}$$

The points inside each cuboid are averaged to obtain one unique point per cuboid, obtaining the following matrix of λ-C_p coordinates:

$$\lambda\text{-}\mathbf{C_p} = \begin{pmatrix} \left(\lambda - C_p\right)_{11} & \cdots & \left(\lambda - C_p\right)_{1n} \\ \vdots & \ddots & \vdots \\ \left(\lambda - C_p\right)_{m1} & \cdots & \left(\lambda - C_p\right)_{mn} \end{pmatrix} \tag{10}$$

To obtain only one λ-C_p point for each λ_j range, a weighted average process is applied considering the m different points contained inside each range. The weighting factor relates the number of samples averaged inside each cuboid to the total number of samples inside each λ_j range, thus the more points inside a cuboid the greater the contribution of that cuboid to the resulting λ-C_p coordinate. The weighting factor w_{ij} applied to each $(\lambda$-$C_p)_{ij}$ point is calculated as in Equation (11).

$$w_{ij} = \frac{N_{ij}}{N_j} \tag{11}$$

where N_j is the total number of points in each λ_j range. **W** is the weights matrix, defined as follows:

$$\mathbf{W} = \begin{pmatrix} w_{11} & \cdots & w_{1n} \\ \vdots & \ddots & \vdots \\ w_{m1} & \cdots & w_{mn} \end{pmatrix} \tag{12}$$

The **B** row vector that contains the λ-C_p points that describe the characteristic curve of the turbines is calculated as follows:

$$\mathbf{B} = W^T \times \left(\lambda - \mathbf{C_p} \right) \tag{13}$$

Figure 6 shows the averaged and weighted points obtained for three aggregation cases: using the data from turbine 1 (blue), merging the data from turbines 1 and 2 (red), and using the combined data from the three turbines (green).

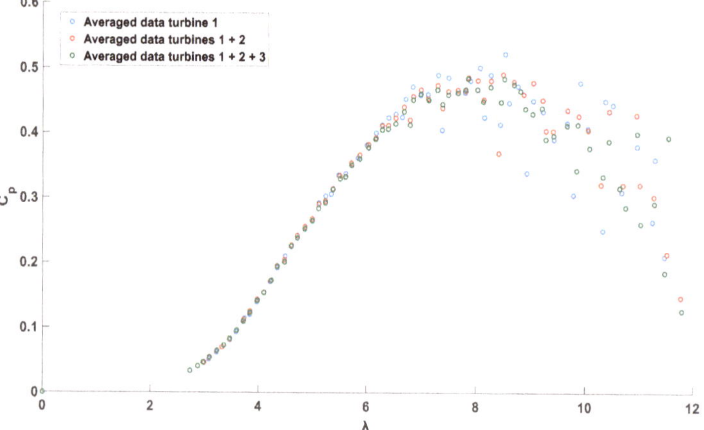

Figure 6. λ-C_p coordinates obtained after the filtering and weighting process.

The resulting λ-C_p points should be then interpolated (step 6 in Figure 3) to obtain the characteristic λ-C_p curve of the turbine. For this purpose, an interpolation method must be selected. To assess the fitting of the estimated curve, the parametric continuity condition is used. In general, a higher order of parametric continuity means a smoother curve with softer gradients. A curve can be said to have C^n parametric continuity when its nth derivative $\left(\frac{d^n y}{dx^n} \right)$ is continuous throughout the curve. Linear interpolation will yield a curve with relatively poor smoothness, usually presenting abrupt gradient changes. The resulting curve of a linear interpolation will have zero-order parametric continuity (denoted as C^0). Polynomial interpolation methods may result in curves with second-order (or higher) parametric continuity, but this method may also suffer from Runge's phenomenon, which appears when applying high-order polynomials to evenly spaced data sets. The cubic spline interpolation method constructs a curve in a piecewise-polynomial fashion. The resulting curve is generated piece-by-piece by applying a low-degree polynomial function for every piece, presenting C^0, C^1 and C^2 parametric continuity. Cubic spline interpolation has a smaller error than linear interpolation and the resulting interpolant function (the λ-C_p curve of the turbine) is easier to evaluate than when polynomial interpolation is used. Due to these characteristics, cubic spline interpolation has been selected as the interpolation method.

The resulting λ-C_p points obtained for this set of turbines are shown as blue dots in Figure 7. It can be seen that the maximum C_p never exceeds 0.5, which is always below the Betz limit. Distribution of points until reaching λ_{opt} is quite uniform, with a greater dispersion for $\lambda > \lambda_{opt}$.

The red curve in Figure 7 is the result obtained after applying the cubic spline interpolation. Figure 7 shows the interpolated points that define the fitting curve along with the averaged points. This figure shows how the spline function, while not being able to pass through every data point, bends the curve to fit the overall trend of the data set. Figure 8 shows the residuals between the interpolated curve and the points used for interpolation. To validate the goodness of fit, some statistical factors have been calculated. The adjusted R-square parameter measures how successful the fit is in explaining the variation of the data. In this case, adjusted R-square of 98.77% implies a very good fit. The root mean square of error is equal to 0.023.

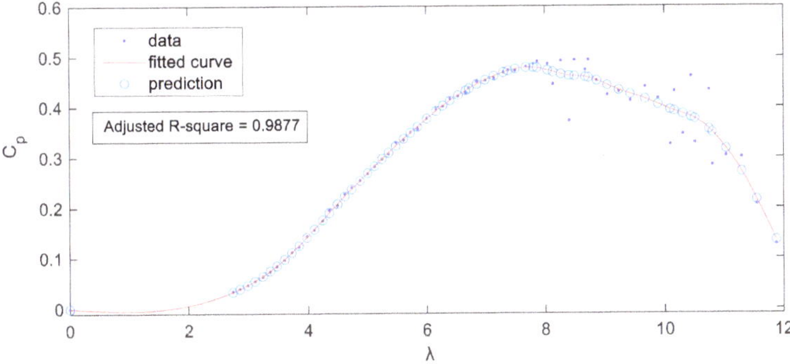

Figure 7. Resulting interpolation from the obtained control points with cubic spline interpolation method.

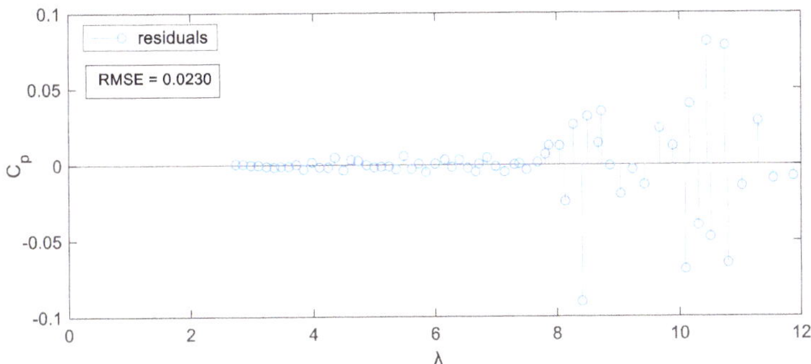

Figure 8. Residuals between the interpolated curve and the points used for interpolation.

The λ-C_p curve estimation improves as more data is fed to the proposed algorithm. Figure 9 shows how the interpolated curve varies as new data is aggregated, showing the curves obtained for the three cases.

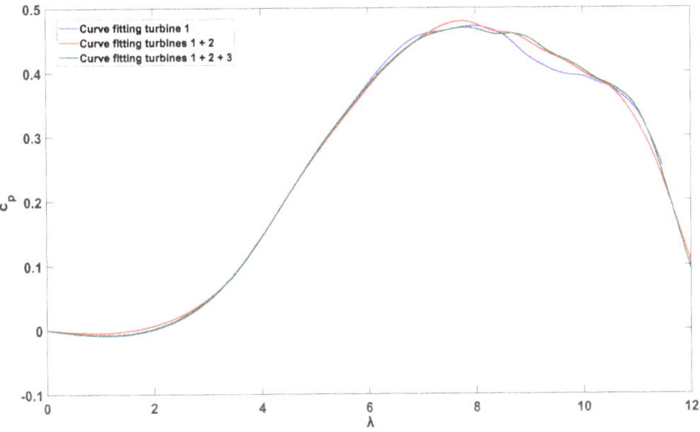

Figure 9. Comparison of the resulting curves as new data is fed into the algorithm.

Analyzing the averaged series, it can be seen that there is no data for λ values below 2.3, so the actual estimated curve starts around $\lambda = 2$. The low-side of the λ-C_p curve (values below $\lambda = 2$) is interpolated from zero to the first available point, keeping the first and second derivatives of the joined curves equal. Although the default value of the first point is $(0, 0)$, it is important to notice that it should be defined by the starting torque of the wind turbine. In Figure 10, the final λ-C_p curve of the turbines is shown.

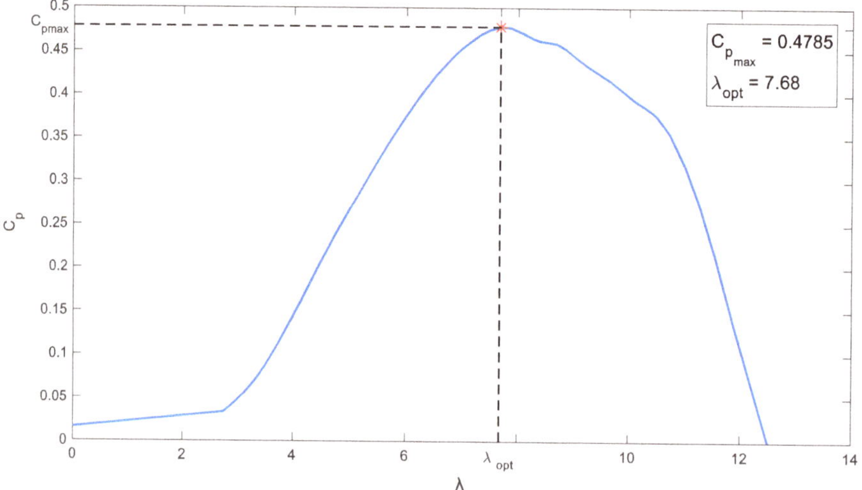

Figure 10. Final λ-C_p curve obtained from three Bornay Wind Plus 25.3+ small wind generators.

The curve reaches a maximum C_p value of 0.4785 at $\lambda = 7.68$. This point represents the maximum efficiency achievable for the system. The resulting curve also considers the starting torque of the real model.

4. Proposed Small Wind Turbine Emulator

The mechanical model of the proposed emulator is shown in Figure 11, where T_T' is the torque delivered by the induction motor that is emulating the turbine operation, N is the gear ratio of the gearbox, and J' is the moment of inertia of the mechanical system including the induction motor (IM block), the gearbox, and the PMSG generator. The emulator is implemented with the same PMSG generator used in the wind turbines used to obtain the λ-C_p curve.

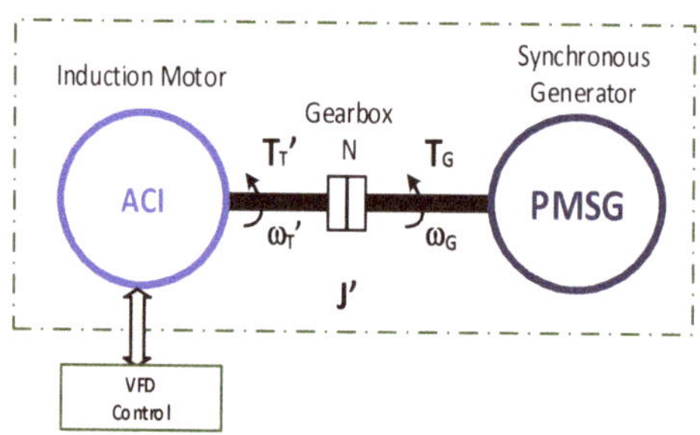

Figure 11. Mechanical model of the proposed wind turbine emulator.

From Figure 11, the rotational dynamics of the proposed wind turbine emulator are expressed as follows:

$$\frac{T_T{}'}{N} - T_G = J'\frac{d\omega_G}{dt} \tag{14}$$

It should be noted that J' must be smaller than J in order to achieve the same dynamic response as the real turbine, compensating the moment of inertia in a similar way to that described in References [17,18]. Basically, emulation of a moment of inertia greater than J' can be achieved by controlling the speed reference variation of the induction motor controller during accelerations or decelerations.

A detailed block diagram of the proposed emulator is shown in Figure 12. A control algorithm that implements Equation (7) determines the control signal applied to the variable-frequency drive (VFD) that, along with the induction motor, simulates the wind turbine. The gearbox enables the induction motor to apply a higher torque to the electric generator side. Therefore, the rated torque of the induction motor might be smaller than the rated torque of the electric generator ($T_T{}' < T_G$) and still be able to emulate the operating point of the turbine on either side of the power curve for testing purposes.

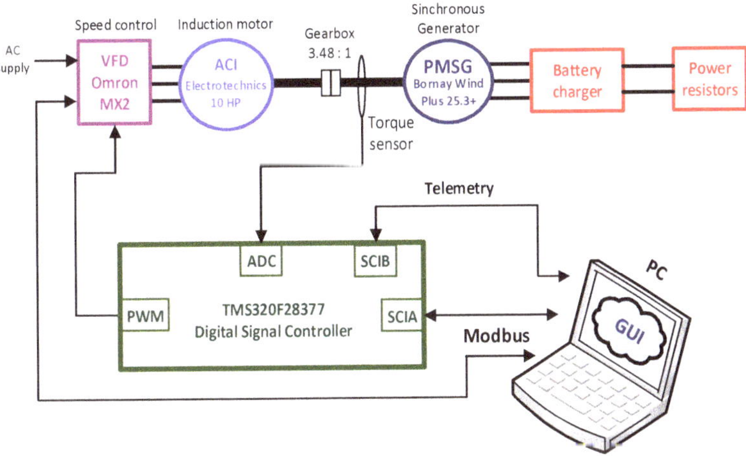

Figure 12. Block diagram of the proposed wind turbine emulator.

The emulated wind torque is provided by an induction motor produced by Electrotechnics (230/400 V/50 Hz, 10 HP, 1460 rpm, $\cos\phi = 0.85$), controlled by an Omron 3G3MX2-A4110-E variable-frequency drive, with the rated characteristics shown in Table 2.

Table 2. Technical characteristics of the variable-frequency drive used in the emulator.

Voltage Class	Constant Torque Control Mode		Variable Torque Control Mode	
	Max Power (kW)	Rated Current (A)	Max Power (kW)	Rated Current (A)
3–400 V$_{L\text{-}L}$	11	24	15	31

The VFD drive implements a sensorless speed control. The speed reference of the VFD drive is proportional to the frequency of a pulse train applied to an input equipped with a capture module. The induction motor and the PMSG are linked by a gearbox with a gear ratio $N = 3.48 : 1$. The PMSG is part of the Bornay Wind Plus 25.3+ small wind turbine and its main characteristics have been described in a previous section. The energy generated by the PMSG must be dissipated by an intelligent load that can be controlled to set different loading situations.

In the proposed emulator, a battery charger and some power resistors implement this intelligent load. The battery charger used in the emulator is part of the solution provided by the manufacturer when the Bornay Wind Plus 25.3+ small wind turbine is used in off-grid systems. Figure 13 details the power vs speed curve used to establish the battery charging current. In this way, for a given rotational speed the battery charger will generate the necessary load torque. The battery charger is equipped with resistors that dissipate the surplus energy and maintain the battery voltage in the floating range, and so avoiding overcharging the battery.

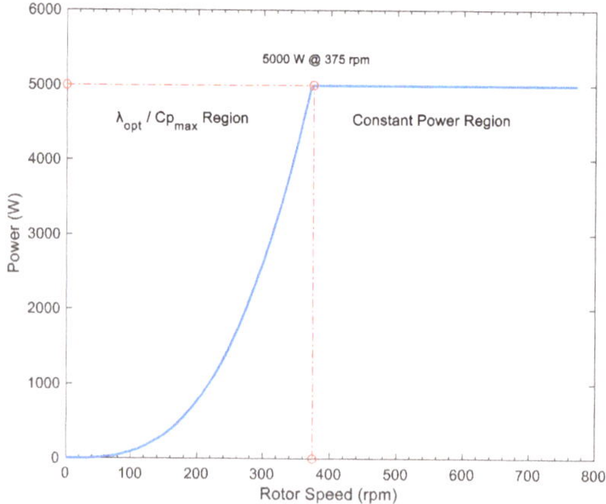

Figure 13. Power-turbine angular speed curve used in the control of the battery charger.

Figure 14 shows the parts of the small wind turbine emulator system: the motor-generator bench in front, followed by the display that shows the GUI (right) and the battery charger (left), and behind is the electric box that contains the VFD drive that controls the induction motors that simulate the turbine (left, front). A gear box connects both shafts.

Figure 14. Experimental set up of the small wind turbine emulator system.

The main control of the turbine emulator system is made by a Texas Instruments TMS320F28377 digital signal controller (DSC). Based on the mechanical model, the inertia to be simulated, and the λ-C_p curve obtained in the previous section, the DSC determines the speed reference of the induction motor. The block diagram of the algorithm implemented in the DSC is depicted in Figure 15. The block diagram implemented corresponds to a directly coupled small wind turbine, so Equation (2) is used and the gear ratio N is added after ω_G is calculated.

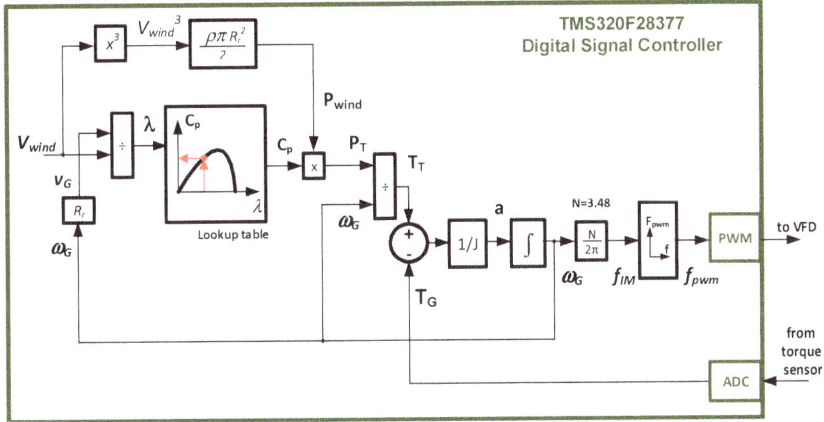

Figure 15. Block diagram of the proposed wind turbine emulator algorithm.

The emulated wind speed is applied to Equation (3) to determine the available wind power (P_{wind}). Wind and turbine speeds are replaced in Equation (6) to obtain the corresponding λ value. At this point, a lookup table with the λ-C_p curve has to be accessed to obtain the actual C_p value. The λ-C_p curve is stored in the memory of the DSC as a table. From these results and by applying Equation (5), the power captured by the turbine (P_T) is computed. The incoming aerodynamic torque (T_T) is obtained by dividing the captured power by the directly coupled turbine speed ($\omega_T = \omega_G$). By calculating the torque difference and using the value of the moment of inertia J to be emulated, the system acceleration ("a" in Figure 15) can then be found from Equation (2). Finally, considering the gearbox ratio, the turbine speed reference signal to feed the speed control input of the VFD drive is obtained by integrating the resulting acceleration.

To validate the proposed small wind turbine emulator system, the experimental λ-C_p curve of Figure 10 has been programmed into the emulator and two experiments have been carried: first, to validate the implementation of the mechanical model, a series of J variations are made; in the second experiment, the emulator is tested along with the GUI.

For the first test, the wind speed is set to a fixed value, battery charger is configured to control the angular speed of the turbine at a set point of 130 rpm, and the speed controller in the battery charger is tuned to control the real turbine. By maintaining the turbine angular speed reference constant, it is verified that the dynamic response of the system varies with the emulated moment of inertia (J), as can be seen for the different cases shown in Figure 16. As J increases, the system response becomes slower, as if the system was getting heavier. This demonstrates that the proposed emulator could emulate the dynamic response of any wind turbine that meets the condition that the moment of inertia of the system to be emulated is greater than the moment of inertia of the emulator.

$$J > J'$$

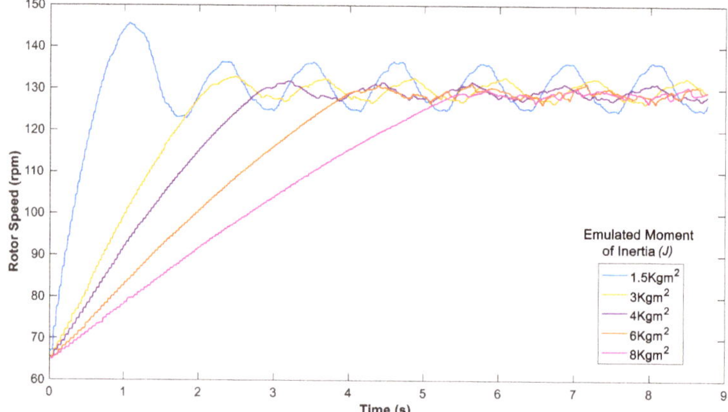

Figure 16. Experimental results of the dynamic of the emulator system for different values of *J'*.

Figure 16 also shows how the response of the speed regulator oscillates accordingly to the value of the moment of inertia. This is because the control of the intelligent load is tuned to work with the Wind Plus 25.3+ turbine ($J = 5.75$ Kg·m^2), as was explained previously. This means that the same controller has a different response when the value of *J* is changed; that is, the load sees a different turbine.

The graphic user interface developed to manage the small wind turbine emulator is shown in Figure 17. Through the GUI, it is possible to configure the VFD drive and communicate with the DSC to start/stop the emulation. In addition, the following emulation parameters can be set inside the GUI environment: rotor radius; height above sea level; speed limit; moment of inertia of the system; wind speed; λ-C_p characteristic curve of the turbine to emulate (left side of the GUI in Figure 17). The GUI also displays a real-time visualization of the operating point of the emulated turbine, both in the λ-C_p curve and in the working surface of the turbine (right section in Figure 17); as well as the emulated wind speed (bottom section in Figure 17).

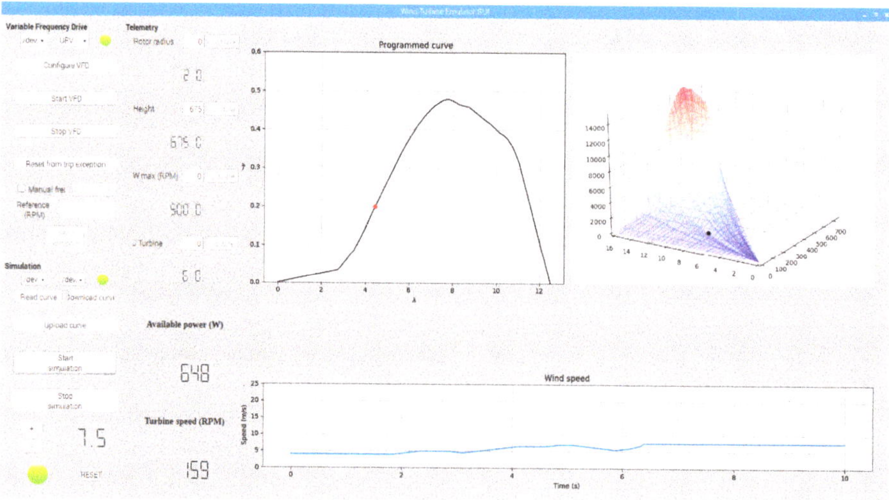

Figure 17. Graphic user interface (GUI) of the wind turbine emulator system. Visualization of the real-time operating point during emulation.

Figure 17 shows the surface of the turbine and the operating point. From the trajectory described by the operating point, the convergence of the models is validated, and the emulator operates at all times on the surface obtained from the experimental data collected. The λ-C_p curve is also shown (upper graph in the center of the GUI screen) along with the real-time C_p value (red dot over the λ-C_p curve), confirming that the emulator presents the same behavior as the real turbine modeled.

The results shown above demonstrate the convergence between the implemented emulator and the model generated from the proposed analysis methodology that is based on the data obtained from three real wind turbines. The experimental λ-C_p curve is uploaded in the GUI and then sent to the DSC that controls the motor/generator bench. The GUI builds and shows the working surface of the corresponding turbine. When the emulator is working, the GUI shows the trajectory of the operating point over the working surface of the turbine. The operation in a stable point located at λ_{opt} demonstrates that the model implemented in the DSC is working correctly and corresponds to that obtained from the real data.

5. Conclusions

The implementation of a realistic small wind turbine emulator system has been addressed in this paper. It is based on the use of the electrical generator and the λ-C_p characteristic curve of the real small wind turbine to be emulated. To validate the proposed emulator, a Bornay Wind Plus 25.3+ wind turbine has been used in this work.

An experimental method for the estimation of the λ-C_p characteristic curve of small wind turbines is proposed. This method avoids the need for expensive wind tunnels; however, a working wind turbine is needed and, depending on the wind conditions, it could take some time to collect enough data to obtain the λ-C_p characteristic curve. A cubic spline interpolation has been used to obtain a smooth λ-C_p curve, which presents a C^2 parametric continuity.

Different aggregations of the data obtained with the three small wind turbines used in the experimental tests are compared. The obtained results demonstrate how the interpolated λ-C_p curve improves when more data sets are used in the process.

A graphic user interface has been developed to configure and monitor the operation of the emulated small wind turbine. It also represents the working surface of the turbine and the real time turbine operating point during the simulation.

The experimental results demonstrate that the dynamic response of the system varies with the emulated moment of inertia, and that the proposed emulator could emulate the dynamic response of any wind turbine with a moment of inertia greater than the moment of inertia of the emulator. The emulator always operates on the working surface obtained from the experimental data collected, demonstrating the convergence between real and emulated systems, showing that the model agrees with that obtained experimentally, and that the emulator works properly.

The proposed emulator will improve the design process and validation of power electronic converters, control strategies, and characterization of the PMSG of small wind turbine systems, by using the real λ-C_p characteristic curve of the wind turbine.

Author Contributions: Conceptualization and methodology, C.I.M.-M., J.D.T.-B. and D.L.-M.; hardware implementation, C.I.M.-M., J.D.T.-B. and F.J.G.-S.; software, C.I.M.-M., D.L.-M. and F.J.G.-S.; results and formal analysis, C.I.M.-M., D.L.-M., F.J.G.-S., S.S.-C. and S.O.-G.; resources and supervision, F.J.G.-S., S.S.-C. and S.O.-G.; writing-review & editing, C.I.M.-M., S.S.-C. and S.O.-G. All authors read and approved the final manuscript.

Funding: This research received no external funding.

Acknowledgments: This work has been carried out as part of a contract with Bornay Aerogeneradores S.L.U., who provided the small wind turbines for the experimental data collection, the mechanical platform of the emulator system, and the technical information of the turbine. Data used to obtain the speed/power curve included in the commercial technical information of the Bornay wind plus 25.3 turbine characterized in this work was obtained from the manufacturer.

Conflicts of Interest: The authors declare no conflicts of interest.

References

1. Nichita, C.; Luca, D.; Dakyo, B.; Ceanga, E. Large band simulation of the wind speed for real time wind turbine simulators. *IEEE Trans. Energy Convers.* **2002**, *17*, 523–529. [CrossRef]
2. Pillay, P.; Krishnan, R. Modeling of Permanent Magnet Motor Drives. *IEEE Trans. Ind. Electron.* **1988**, *35*, 537–541. [CrossRef]
3. Carrillo Arroyo, E.L. Modeling and Simulation of Permanent Magnet Synchronous Motor Drive System. Master's Thesis, University of Puerto Rico, San Juan, Puerto Rico, 2006.
4. Tarımer, İ.; Ocak, C. Performance Comparision of Internal and External Rotor Structured Wind Generators Mounted from Same Permanent Magnets on Same Geometry. *Elektron. IR Elektrotechnika* **2009**, *92*, 65–70.
5. Morales, G.; Gerardo, L. Mejora de la Eficiencia y de las Prestaciones Dinámicas en Procesadores Electrónicos de Potencia Para Pequeños Aerogeneradores Sincrónicos Operando en Régimen de Velocidad Variable. Ph.D. Thesis, Universitat Politècnica de València, Valencia, Spain, May 2011.
6. Tanvir, A.; Merabet, A.; Beguenane, R. Real-Time Control of Active and Reactive Power for Doubly Fed Induction Generator (DFIG)-Based Wind Energy Conversion System. *Energies* **2015**, *8*, 10389–10408. [CrossRef]
7. Li, B.; Tang, W.; Xiahou, K.; Wu, Q. Development of Novel Robust Regulator for Maximum Wind Energy Extraction Based upon Perturbation and Observation. *Energies* **2017**, *10*, 569.
8. Escalante, E.; Castellanos, R.; López, O. Diseño e implementación de un control de máxima potencia para un sistema de generación de energía eólica con generador síncrono de imán permanente. Design and implementation of a maximum power control of a wind power generation system with a permanent mag. *Ingenieria* **2012**, *16*, 1–20.
9. Diaz, S.A.; Silva, C.; Juliet, J.; Miranda, H.A. Indirect sensorless speed control of a PMSG for wind application. In Proceedings of the 2009 IEEE International Electric Machines and Drives Conference, Miami, FL, USA, 3–6 May 2009; pp. 1844–1850.
10. Xue, X.; Bu, Y.; Lu, B. Development of a nonlinear wind-turbine simulator for LPV control design. In Proceedings of the 2015 IEEE Green Energy and Systems Conference (IGESC), Long Beach, CA, USA, 9 November 2015; pp. 41–48.
11. Martinez, F.; Herrero, L.C.; Pablo, S.D. Open loop wind turbine emulator. *Renew. Energy* **2014**, *63*, 212–221.
12. Castelló, J.; Espí, J.M.; García-Gil, R. Development details and performance assessment of a Wind Turbine Emulator. *Renew. Energy* **2016**, *86*, 848–857. [CrossRef]
13. Liu, B.; Nishikata, S.; Tatsuta, F.; Suzuki, K. A wind turbine simulator considering various moments of inertia using a DC motor. In Proceedings of the 2014 International Symposium on Power Electronics, Electrical Drives, Automation and Motion, Ischia, Italy, 18–20 June 2014; pp. 866–870.
14. Liu, B.; Tatsuta, F.; Nishikata, S.; Suzuki, K. A study on a wind turbine simulator with a DC motor considering various moments of inertia. In Proceedings of the in 2014 17th International Conference on Electrical Machines and Systems (ICEMS), Hangzhou, China, 22–25 October 2014; pp. 3010–3015.
15. Hussain, J.; Mishra, M.K. Design and development of real-time small scale wind turbine simulator. In Proceedings of the 2014 IEEE 6th India International Conference on Power Electronics (IICPE), Kurukshetra, India, 8–10 December 2014; pp. 1–5.
16. Kojabadi, H.M.; Chang, L.; Boutot, T. Development of a Novel Wind Turbine Simulator for Wind Energy Conversion Systems Using an Inverter-Controlled Induction Motor. *IEEE Trans. Energy Convers.* **2004**, *19*, 547–552. [CrossRef]
17. Weijie, L.; Minghui, Y.; Rui, Z.; Minghe, J.; Yun, Z. Investigating instability of the wind turbine simulator with the conventional inertia emulation scheme. In Proceedings of the 2015 IEEE Energy Conversion Congress and Exposition (ECCE), Montreal, QC, Canada, 20–24 September 2015; pp. 983–989.
18. Gong, B.; Xu, D. Real time wind turbine simulator for wind energy conversion system. In Proceedings of the 2008 IEEE Power Electronics Specialists Conference, Rhodes, Greece, 15–19 June 2008; pp. 1110–1114.
19. Choy, Y.-D.; Han, B.-M.; Lee, J.-Y.; Jang, G.-S. Real-Time Hardware Simulator for Grid-Tied PMSG Wind Power System. *J. Electr. Eng. Technol.* **2011**, *6*, 375–383. [CrossRef]
20. Dolan, D.S.L.; Lehn, P.W. Real-time wind turbine emulator suitable for power quality and dynamic control studies. In Proceedings of the International Conference on Power Systems Transients (IPST'05), Montreal, QC, Canada, 19–23 June 2005.

21. Cutululis, N.; Ciobotaru, M.; Ceanga, E.; Rosu, M.E. Real Time Wind Turbine Simulator Based on Frequency Controlled AC Servomotor. *Ann. Dunarea Jos Univ. Galati* **2002**, *3*, 97–101.

22. Wasynczuk, O.; Man, D.T.; Sullivan, J.P. Dynamic Behavior of a Class of Wind Turbine Generators during Random Wind Fluctuations. *IEEE Power Eng. Rev.* **1981**, *PER—1*, 47–48. [CrossRef]

23. Dai, J.; Liu, D.; Wen, L.; Long, X. Research on power coefficient of wind turbines based on SCADA data. *Renew. Energy* **2016**, *86*, 206–215. [CrossRef]

Article

Consensus-Based SOC Balancing of Battery Energy Storage Systems in Wind Farm

Cao-Khang Nguyen [1], Thai-Thanh Nguyen [1], Hyeong-Jun Yoo [1] and Hak-Man Kim [1,2,*]

1   Department of Electrical Engineering, Incheon National University, Songdo-dong, 119 Academy-ro,
    Yeonsu-gu, Incheon 22012, Korea; khang.nguyencao@gmail.com (C.-K.N.); ntthanh@inu.ac.kr (T.-T.N.);
    yoohj@inu.ac.kr (H.-J.Y.)
2   Research Institute for Northeast Asian Super Grid, Incheon National University, 119 Academy-ro,
    Yeonsu-gu, Incheon 22012, Korea
*   Correspondence: hmkim@inu.ac.kr; Tel.: +82-32-835-8769; Fax: +82-32-835-0773

Received: 17 November 2018; Accepted: 13 December 2018; Published: 16 December 2018

Abstract: Multiple battery energy storage systems (BESSs) are used to compensate for the fluctuation in wind generations effectively. The stage of charge (SOC) of BESSs might be unbalanced due to the difference of wind speed, initial SOCs, line impedances and capabilities of BESSs, which have a negative impact on the operation of the wind farm. This paper proposes a distributed control of the wind energy conversion system (WECS) based on dynamic average consensus algorithm to balance the SOC of the BESSs in a wind farm. There are three controllers in the WECS with integrated BESS, including a machine-side controller (MSC), the grid-side controller (GSC) and battery-side controller (BSC). The MSC regulates the generator speed to capture maximum wind power. Since the BSC maintains the DC link voltage of the back-to-back (BTB) converter that is used in the WECS, an improved virtual synchronous generator (VSG) based on consensus algorithm is used for the GSC to control the output power of the WECS. The functionalities of the improved VSG are designed to compensate for the wind power fluctuation and imbalance of SOC among BESSs. The average value of SOCs obtained by the dynamic consensus algorithm is used to adjust the wind power output for balancing the SOC of batteries. With the proposed controller, the fluctuation in the output power of wind generation is reduced, and the SOCs of BESSs are maintained equally. The effectiveness of the proposed control strategy is validated through the simulation by using a MATLAB/Simulink environment.

Keywords: wind energy conversion system; distributed control; battery energy storage system; consensus algorithm

1. Introduction

Renewable energy sources (RESs) such as wind power and solar energy have been widely integrated into the power system [1–3]. One of the fastest growing RESs is wind power, which has received more and more attention from researchers in recent years [4]. However, the uncertainty of wind power results in the fluctuation in system frequency. Therefore, an increase in the penetration of wind power might cause several adverse effects to the power quality, stability and reliability of the power system [5,6]. In order to mitigate the impacts of wind power fluctuation, battery energy storage system (BESS) has been integrated into the wind energy conversion system (WECS) [7,8]. By controlling the charging/discharging power of the battery, the output power of the WECS can be smoother. For the large-scale wind farm, multiple BESSs could be used for addressing the problem of wind power fluctuation [9].

The imbalance of state of charge (SOC) is a common issue in the multi-cell battery systems due to the manufacturing variance, internal impedance, self-discharge rates, etc. [10,11]. The SOC imbalance might result in the reduction of charge capacity, early termination of charging or discharging, and accelerated battery degradation [10]. This problem is also observed in the microgrid or wind farm systems with multiple BESSs that are operated independently. Several studies have addressed the issue of SOC imbalance in the independent BESSs [12–22]. The SOC of the BESSs is unbalanced due to the difference of wind power output, initial SOC, capacities of BESSs, and line impedances [12]. The batteries with the lowest SOC might stop the operation of the WECS because it does not have enough power to support the system. The remaining batteries should be in charge of compensating power for the load change, which results in the accelerated charge or discharge rate of BESSs and the reduction of service life of BESSs [13]. The system could be collapsed if the energy storage of the remaining BESSs is not enough to compensate for the disturbance [14]. When multiple BESSs are used in the wind farm, the system reliability can be improved by maintaining the state of charges (SOC) balancing of BESSs [12].

In order to deal with the unbalanced SOC problem, several control strategies for SOC balancing have been introduced [15–22]. The SOC-based droop control strategies of the converter have been used widely to achieve proper power sharing and SOC balancing. The authors in reference [15] proposed a double-quadrant SOC-based droop control to achieve an appropriate power sharing in the autonomous DC microgrid. The SOC-based adaptive droop control was introduced for balancing the SOCs among the energy storage systems (ESU) in DC microgrid [16]. The paper in [17] proposed a multifunctional and wireless droop control to eliminate the SOC imbalance of the ESUs based on the $P - f$ droop controller. However, the SOC-based droop control strategies cause the frequency deviation from the nominal values, which affects to the performance of the system.

Instead of using the SOC-based droop control, several SOC balancing control schemes based on the centralized control were proposed to balance the discharge rate of the energy storage system [18,19]. A coordinated secondary control based on current sharing control was introduced to achieve the discharge rate balancing and avoid the overcurrent among distributed generations in the islanded AC microgrid. However, the use of the centralized controller for SOC balancing might reduce the system reliability since all converters rely on the centralized controller [20]. The SOC balancing controllers based on decentralized control or distributed control have been proposed recently to improve the problem of SOC imbalance and enhance the system reliability [21,22]. In reference [21], a multi-agent based SOC balancing control was discussed, while in reference [22], the SOC balancing control was achieved by using coordinated secondary control to adjust virtual impedance loops.

However, most of the papers in the literature mainly focused on the stand-alone system. In addition, the impact of wind generation on the SOC of batteries was neglected. In this paper, an improved virtual synchronous generator (VSG) control based on consensus algorithm is proposed to regulate the output power of wind energy conversion system, which considers the impacts in wind speed variation on the WECS. The proposed controller consists of two main parts, including a consensus based SOC balancing controller and smoothing power controller. With the use of consensus-based control algorithm, the SOC of batteries is controlled to converge at the same value. The batteries which have higher SOC will inject more power to support the batteries with lower SOC. In addition, the low-pass filter is added to the VSG control of grid-side converter for smoothing active output power of the WECS. Hence, the performance of the WECS can be improved, and the system stability can be enhanced.

The paper is arranged as follows: Section 2 presents the overall structure of the WECS and the detailed control diagram of power converters in the WECS. The proposed control strategy based on consensus algorithm is introduced in Section 3. The performance of the proposed control strategy is validated through the simulation results in Section 4. Finally, the conclusion appears in Section 5.

2. Structure of Wind Energy Conversion System with Integrated Battery Energy Storage System

2.1. Overall Structure

The overall structure of the WECS with integrated BESS is illustrated in Figure 1. The wind turbine is directly connected to the permanent magnet synchronous generator (PMSG). The PMSG is interfaced with the utility grid through the back-to-back (BTB) converter, which includes three main parts, namely, machine-side converter (MSC), grid-side converter (GSC) and battery-side converter (BSC). The MSC controls the generator in order to capture maximum active power from wind turbine by using maximum power point tracking (MPPT) algorithm. In the BSC, a bidirectional DC-DC converter is utilized to maintain the constant DC capacitor voltage of the BTB converter by charging or discharging power. With the use of BESS, instead of controlling the DC capacitor voltage, the GSC is flexible to adjust the output power and support the system frequency. The power converter used in the GSC can be classified into two types, grid-feeding converter and grid-forming converter [23]. In this structure, the grid-forming converter is adopted to regulate the active power and support the system frequency by using the VSG control. The advantages of VSG control for supporting system frequency by imitating both steady-state and transient-state characteristics of the synchronous generator are explicitly discussed in previous works [24,25]. Therefore, it will not be further considered in this paper. The detailed control diagram of each converter in the WECS is analyzed in the next part.

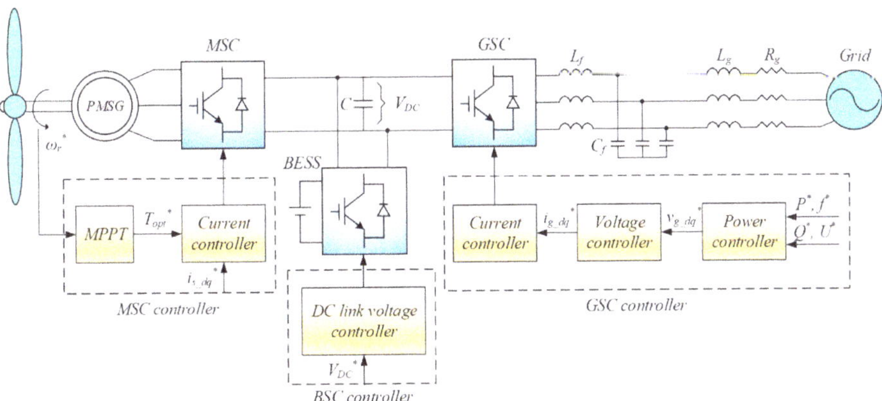

Figure 1. Overall structure of the wind energy conversion system (WECS) integrated with battery energy storage system (BESS).

2.2. Control of a Machine-Side Converter

The machine-side converter is utilized to control the generator of the WECS. Depending on the variation of wind speed, the rotor speed is varied, and thus, the output power generated by the PMSG fluctuates. By applying the MPPT algorithm, electromagnetic torque is regulated to be proportional to the square of rotor speed so that the optimal value is maintained. Therefore, with particular wind speed, the wind turbine can produce maximum possible power. The control scheme of MSC is shown in Figure 2a, which includes two cascaded controllers, namely torque controller, and current controller. With regard to the torque controller, the reference of electromagnetic torque is achieved by the MPPT algorithm, as shown below:

$$T_e^* = -k_{opt}\omega_r^2 \tag{1}$$

where T_e^* is the optimal torque generated from wind turbine, ω_r is the speed of rotor, k_{opt} is the optimal coefficient wind turbine, which is described as:

$$k_{opt} = \frac{0.5\rho A r^3 C_{pmax}}{\lambda_{opt}^3}\omega_r^3 \tag{2}$$

where A is the turbine swept area, r is the turbine radius, ρ is the air mass density, λ_{opt} is the optimal tip-speed ratio when the blade pitch angle $\beta = 0$, C_{pmax} is the maximum performance coefficient.

The optimal torque determined by the torque controller is used to calculate the current reference of the current controller, as given by:

$$i_{sq}^* = \frac{2T_e^*}{3p\lambda_m} \tag{3}$$

where λ_m is the flux generated from the permanent magnet, i_{sq}^* is the reference of stator current in q-axis, p is the number of pole pairs in the PMSG.

While the MPPT algorithm is applied to the torque control loop, the current controller adjusts the stator current of the generator in dq reference frame. The current in d-axis is controlled at zero to maintain the linearization between electrical torque and the q-axis current, as shown in Equation (3). The output voltage reference of the generator can be determined through the inner current loop, which is given by:

$$v_d^* = k_{pm}(i_{sd}^* - i_{sd}) + k_{im} \int (i_{sd}^* - i_{sd})dt - \omega_r L_q i_{sq} \tag{4}$$

$$v_q^* = k_{pm}\left(i_{sq}^* - i_{sq}\right) + k_{im} \int \left(i_{sq}^* - i_{sq}\right)dt + \omega_r L_d i_{sd} + \omega_r \lambda_m \tag{5}$$

where i_{sd} and i_{sq} are the stator current in dq reference frame, v_d^* and v_q^* are the reference of stator voltage of the PMSG and k_{pm} and k_{im} are the PI parameters of the current controller in the MSC.

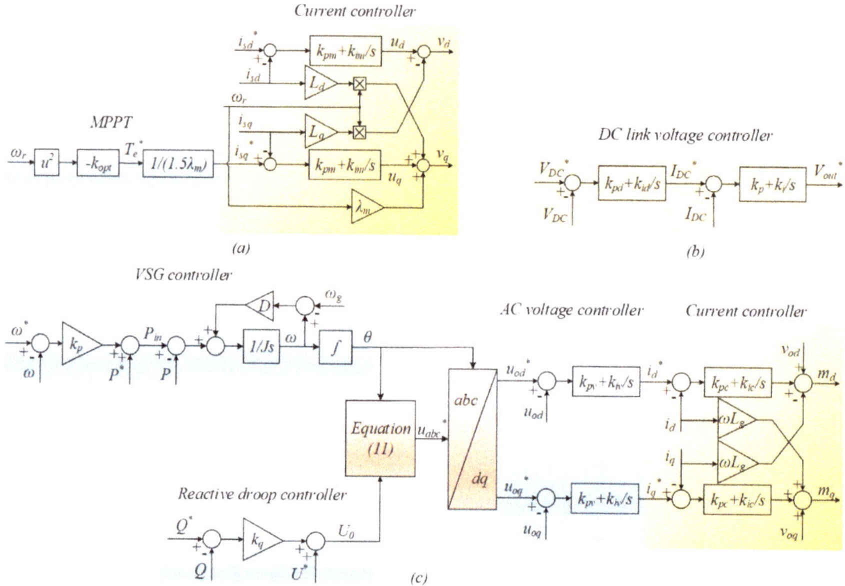

Figure 2. Control diagram of power converter in the WECS: (**a**) Machine-side converter, (**b**) Battery-side converter, (**c**) Grid-side converter.

2.3. Control of a Battery-Side Converter

The BSC adopts the bi-directional DC-DC converter to regulate the DC capacitor voltage of the BTB converter. In addition, it plays an important role in balancing the power between the GSC and the MSC. When the wind power is lower than the power required from the grid, the power from the BSC is released. By contrast, when the power generated from a wind turbine is larger than the power requirement from the grid, the power is reserved by the BSC. The detailed control scheme of the BSC is shown in Figure 2b. Whereas the outer control is designed to maintain the constant DC capacitor voltage of the BTB converter, the inner controller adjusts the current flowing to the inductor of the bi-directional DC-DC converter. The output voltage and current controller of the converter can be calculated by these equations:

$$I_{DC}^* = k_{pd}(V_{DC}^* - V_{DC}) + k_{id} \int (V_{DC}^* - V_{DC})dt \tag{6}$$

$$V_{out}^* = k_p(I_{DC}^* - I_{DC}) + k_i \int (I_{DC}^* - I_{DC})dt \tag{7}$$

where k_{pd} and k_{id} are the PI parameters of the DC capacitor voltage controller, k_p and k_i are the PI parameters of the current controller in the bidirectional DC-DC converter and V_{out}^* is the output voltage reference.

The design of the PI parameters of voltage and current controller should consider the different bandwidth in each control loop. The current controller is designed for a bandwidth of 1.6 kHz with good rejection of high-frequency disturbance, whereas the voltage controller is designed for 400 Hz bandwidth to ensure the stability of the converter system [26]. The classical pole-zero and bode techniques are used to select these parameters.

2.4. Control of a Grid-Side Converter

With regard to the grid-side converter, its detailed control diagram is shown in Figure 2c, which consists of three cascaded control loops: Power controller, voltage controller and current controller. In the power controller, virtual synchronous generator control is applied to active power control to regulate the power angle θ while reactive power control adjusts the AC voltage reference by using droop controller. With the use of VSG control, the GSC can imitate both steady-state and transient-state characteristics of a synchronous generator to virtually provide inertia to the WECS. The VSG control is described by the swing equation, which is given by:

$$P_{in} - P = J\omega\frac{d\omega}{dt} + D(\omega - \omega_g) \tag{8}$$

where P is the measured active power of the GSC, J is the virtual inertia coefficient, ω is the rotor angular frequency generated by VSG control, ω_g is the grid frequency and P_{in} is the input power which is determined by the governor:

$$P_{in} = P^* + k_p(\omega^* - \omega) \tag{9}$$

where P^* the set active power, k_p is the droop coefficient for active power controller and ω^* is the nominal angular frequency.

Regarding the reactive power controller, the amplitude of the output voltage reference in *abc* reference frame can be calculated by the droop controller, which is given as:

$$U_0 = U^* + k_q(Q^* - Q) \tag{10}$$

where U_0 is the amplitude of output AC voltage, k_q is the droop coefficient for reactive power controller and Q^* is the set reactive power.

The reference of output voltage for the GSC is determined by combining Equations (9) and (10), which is shown as follows:

$$\begin{cases} u_a^* = U_0\sin\theta \\ u_b^* = U_0\sin(\theta - 2\pi/3) \\ u_c^* = U_0\sin(\theta + 2\pi/3) \end{cases} \tag{11}$$

The reference of the output AC voltage is transferred from *abc* reference frame to *dq* reference frame, and then input to the inner control loop. The inner control loop consists of two controllers, namely voltage controller and current controller. The voltage controller is used to maintain the output voltage of the GSC, which is described as follows:

$$i_{ld}^* = k_{pv}(v_{od}^* - v_{od}) + k_{iv}\int (v_{od}^* - v_{od})dt \tag{12}$$

$$i_{lq}^* = k_{pv}\left(v_{oq}^* - v_{oq}\right) + k_{iv}\int \left(v_{oq}^* - v_{oq}\right)dt \tag{13}$$

where k_{pv} and k_{iv} are the proportional and integral gain of voltage controller, respectively.

Regarding the current controller, it is employed to control the inductor current of the GSC. The current reference is also the output of the voltage controller. Similar to the voltage controller, the reference value of the current controller is also compared with the measured current and then input to the PI controller. The output of the current controller can be treated by:

$$m_d = v_{od} - \omega i_q L_f + k_{pc}(i_{ld}^* - i_{ld}) + k_{ic}\int (i_{ld}^* - i_{ld})dt \tag{14}$$

$$m_q = v_{oq} + \omega i_d L_f + k_{pc}\left(i_{lq}^* - i_{lq}\right) + k_{ic}\int \left(i_{lq}^* - i_{lq}\right)dt \tag{15}$$

where k_{iv} and k_{iv} are the proportional and integral component of the current controller, respectively, and m_d and m_q are the modulating signals in the *dq* reference frame.

3. Proposed Control Strategy

In the proposed control method, a consensus algorithm is applied for achieving SOC balance in the WECS while the low pass filter is added to compensate the power fluctuation of the WECS. The average dynamic consensus algorithm was introduced in reference [27], which is presented by:

$$\dot{\zeta}_i = \sum_{j=1}^{N} a_{ij}\left(\zeta_i - \zeta_j\right) \qquad i = 1, 2, \ldots, N \tag{16}$$

where ζ_i and ζ_j are the state variables for node i and j, respectively, a_{ij} is the weight for exchanging information from node j to i.

The communication network is described by a graph with nodes N and edges . Each node of the graph represents an agent and each edge of edges represents a communication link between two particular agents. The agents communicate with their neighbors to exchange information among them, and then update the state information. The Laplacian matrix $L_N = [l_{ij}]\epsilon R^{n\times n}$ associated with the graph can be written as:

$$\begin{cases} l_{ii} = \sum_{j=1, \ j\neq i}^{N} a_{ij} \\ l_{ij} = -a_{ij} \ j \neq i \end{cases} \tag{17}$$

Therefore, the dynamics of the consensus algorithm can be described by:

$$\dot{X} = -L_N X \tag{18}$$

where $X = [x_1, x_2, \ldots, x_n]^T$ is the state variable vector.

The SOC of batteries among WECSs are considered as the state variables of the consensus algorithm. Through the consensus algorithm, the reference of SOC is updated after each time interval. The current SOC of battery i can be calculated as:

$$SOC_i = SOC_i^* - \frac{1}{C_{bati}} \int_0^t I_i dt \tag{19}$$

where SOC_i^* and SOC_i are the initial SOC and the current state of charge of battery i, I_i and C_{bati} are the output current and the capacity of battery i.

The power generated from the SOC controller can be calculated by:

$$P_{SOC} = k_{SOC}(SOC^* - SOC_i) = k_{SOC}\Delta SOC_i \tag{20}$$

where k_{SOC} is the SOC coefficient, SOC^* is the SOC reference updated from the consensus algorithm and ΔSOC_i is the difference between the updated SOC reference and the current SOC of battery i.

A low-pass filter is used to smoothen the fluctuation of the output wind power, and then the subtraction of filtered wind power from the measured wind power is used as the compensated power for the fluctuation. Since each WECS has the ability to smoothen wind power, the total power of all WECSs transferred to the grid could be smoothed. The reference of output power is added to regulate the output power of the WECS, as shown in Figure 3.

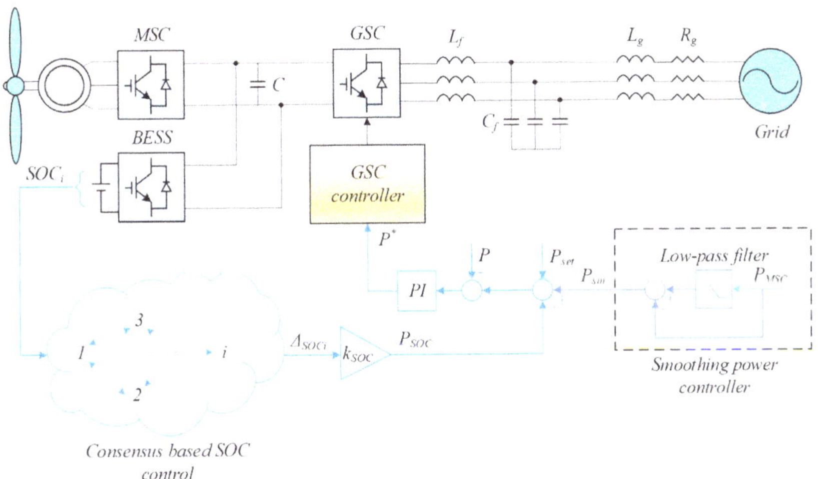

Figure 3. Control diagram of the proposed control strategy.

4. Simulation Results

In order to evaluate the efficiency of the proposed control strategy, a simulation system is carried out by using a SimPowerSystems toolbox in a MATLAB/Simulink environment. The detailed simulation system which consists of three WECSs with integrated BESS is presented in Figure 4. These WECSs are connected in parallel through the impedance lines and then interfaced with the utility grid. Each WECS injects 1.5 (MW) active power to the utility grid. The proposed controller is verified through two case studies. Whereas the performance of the system is analyzed with different initial SOCs in case study 1, the same initial SOC is focused on case study 2. The simulation parameters of PMSG are listed in Table 1.

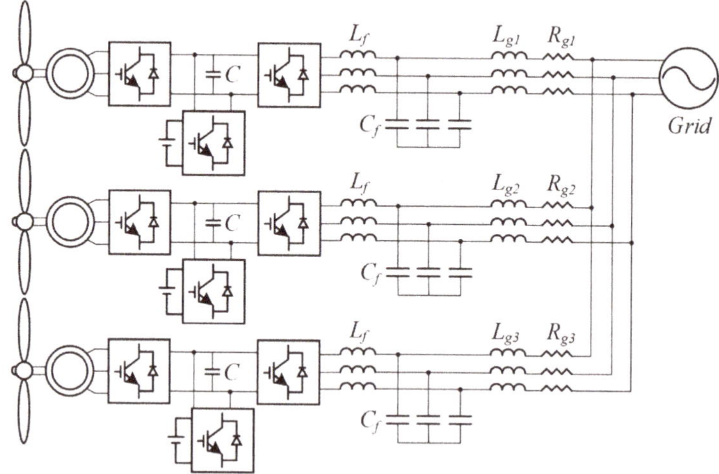

Figure 4. Simulation system.

Table 1. System parameters.

Parameters	Values	Parameters	Values
Rated power	2.45 (MW)	Number of pole pairs	8
Rated line-to-line voltage	4000 (V)	Synchronous inductance	9.816 (mH)
Rated stator current	490 (A)	Rated rotor speed	400 (rpm)
Rated stator frequency	53.3 (Hz)	Flux	7.03 (Wb)
Active power droop coefficient	4×10^{-5}	Reactive power droop coefficient	0.068

4.1. Case Study 1: Performance of the Proposed Control Strategy with Different Initial SOCs

In the first case, the average SOC of batteries is chosen to equal 45%, 35%, and 25% at the beginning, respectively. The wind speed flowing to three wind turbines is chosen differently for each one, as shown in Figure 5. Whereas the average wind speed of WECS1 is the highest, WECS3 receives the lowest average wind speed. The variation of wind speed results in the fluctuation of active power generated from the wind turbine, which is measured through the MSC, as shown in Figure 6. Although the output power transferred to the utility grid is regulated at 1.5 (MW) for each WECS, the power captured from the wind turbine is not enough for satisfying the power requirement of the utility grid. Therefore, the batteries discharge power in order to achieve power balance between the MSC and the GSC and compensate the power fluctuation.

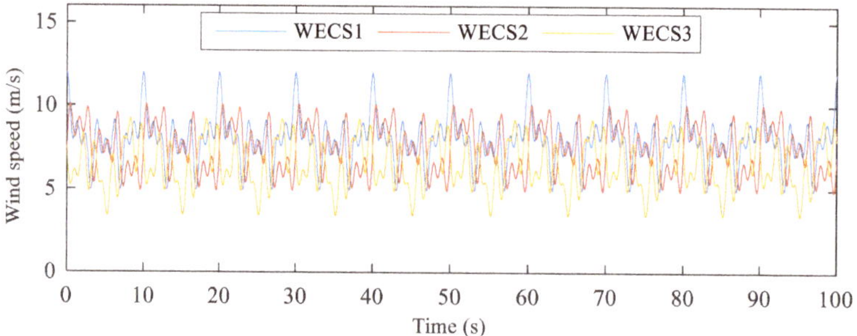

Figure 5. Wind speed.

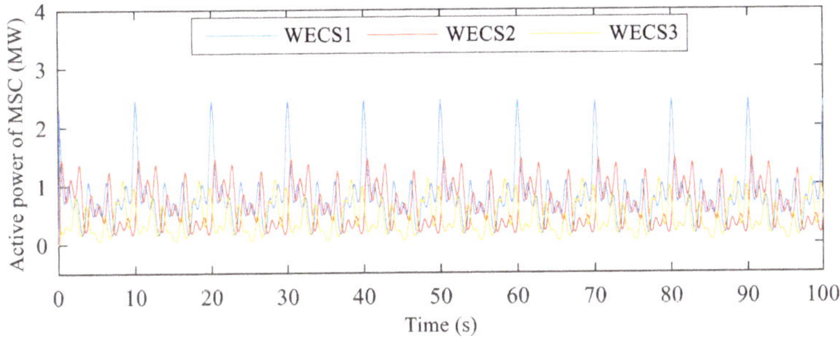

Figure 6. Active power of machine-side converter.

It can be seen from Figure 7 that, without the proposed control strategy, the SOC of each battery moves down slightly at the first interval of 50 s. After that, whereas the SOC of battery 1 and battery 2 are higher than 20%, the SOC of battery 3 moves down to under 20%. However, when the consensus algorithm is applied, the SOC of battery 1 moves down quickly while that of battery 3 moves up in the first 30 s, as shown in Figure 8. In other words, battery 1, which has the highest SOC, releases more power to support the other batteries. Consequently, battery 2 discharges less power and battery 3 reserves power because it has the lowest SOC. After 50 s, all three batteries discharge power to the GSC with the same SOC. In addition, none of the batteries have SOC below 20%. Hence, the proposed controller can reduce the adverse effects of the low SOC on the DC capacitor voltage and the WECS, which has the lowest SOC, can survive.

Furthermore, Figure 9 presents the active power of the GSC without the proposed controller. Because the SOC of batteries is not regulated, the active output power of the GSC is maintained equally at 1.5 MW. However, as illustrated in Figure 10, when the proposed controller is applied, WECS1, which has the highest SOC, injects more power than WECS2 and WECS3 until the SOCs among the WECSs reach the same value at roughly 50 s. After the SOCs are balanced, the output power exchanged with the utility grid is maintained at the reference value. In addition, with the use of smoothing power controller, the fluctuation of wind power is compensated by the battery of each WECS, as shown in Figures 11 and 12. Therefore, the output power of WECSs can be smooth, which enhances the power quality of the system.

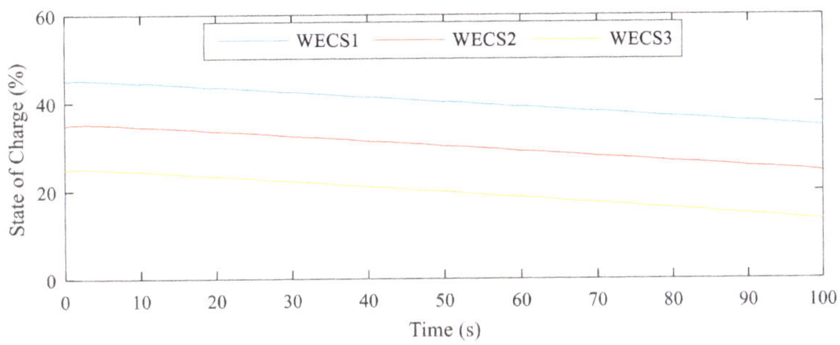

Figure 7. State of charge (SOC) of batteries in the WECS without proposed controller (Case 1).

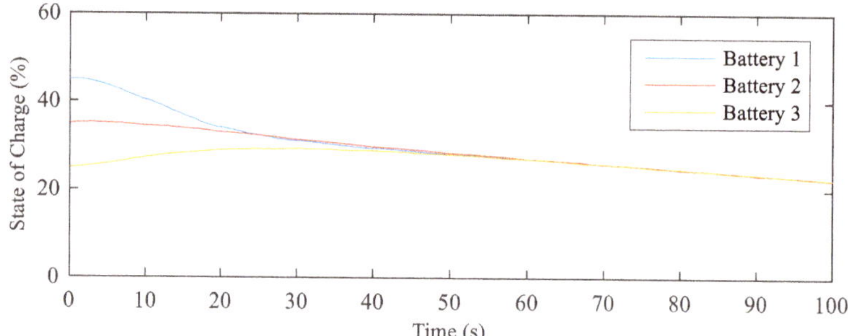

Figure 8. State of charge of batteries in the WECS with proposed controller (Case 1).

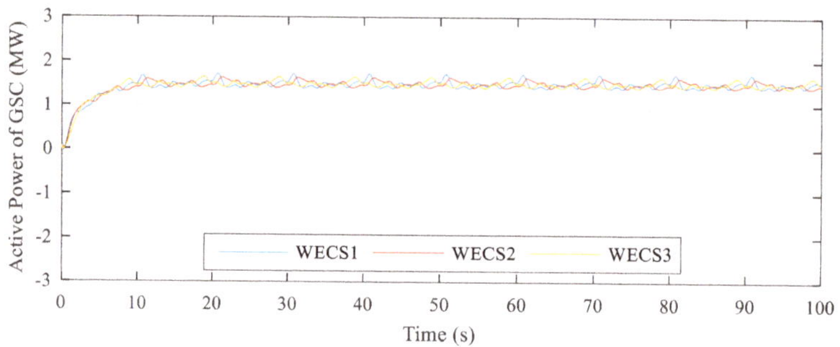

Figure 9. Active power of grid-side converter (GSC) without the proposed controller (Case 1).

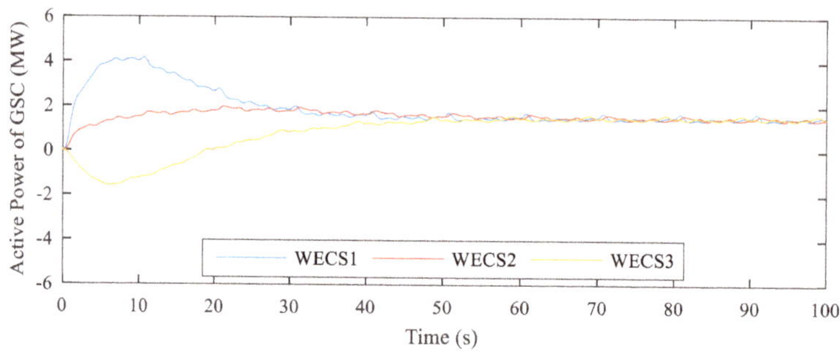

Figure 10. Active power of GSC with the proposed controller (Case 1).

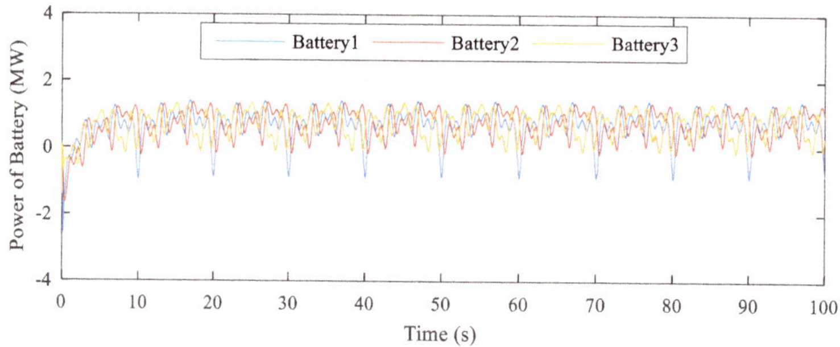

Figure 11. Power of battery without the proposed controller (Case 1).

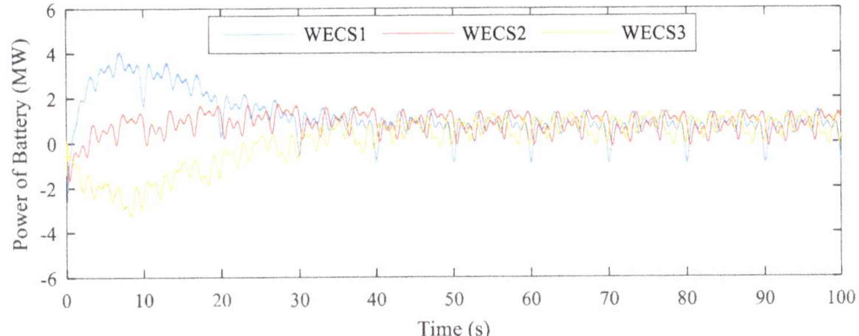

Figure 12. Power of battery with the proposed controller (Case 1).

4.2. Case Study 2: Performance of the Proposed Control Strategy with Same Initial SOCs

In the second case study, the effectiveness of the proposed controller is verified when the initial SOC of batteries is designed to have the same values. As can be seen from Figure 13, the SOC of batteries remains 50% at the beginning. Without the proposed controller, due to the different wind speeds of the WECSs, the SOCs slightly move far away from each other at the first interval of 20 s.

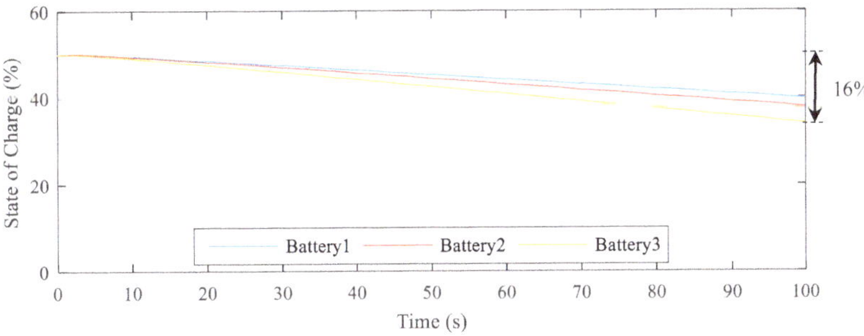

Figure 13. State of charge of batteries in the WECS without proposed controller (Case 2).

After that, it can be clearly seen that the SOCs are significantly different over the 100-s period. The SOC of battery 3 moves down quickly, and its slope is higher in comparison with other SOCs because battery 3 releases power the most. It reaches 34% at the end of the period, corresponding to 16% of power discharge.

On the other hand, Figure 14 shows the SOC of the batteries in case of applying consensus algorithm. From 50% at the initial time, the SOC of batteries moves down with a moderate slope, and they remain at a similar value until the end of the period. Moreover, the lowest amount of the SOC is 37%, corresponding to 13% of the power discharge. This value is smaller than the SOC of battery 3 without the consensus algorithm. Therefore, battery 3 releases less power to the GSC in this case.

In addition, Figures 15 and 16 illustrate the active output power flowing to the utility grid without the consensus algorithm and with the consensus algorithm, respectively. It is clear that in both cases the active power of GSCs is controlled to maintain at 1.5 MW. The power fluctuation is reduced by using a low-pass filter to damp out the high-frequency oscillation. For that reason, the active power of GSCs has small variation. In addition, without the proposed controller, the active power of GSCs are injected equally to the grid, while they have a small difference when the consensus algorithm is applied. It is because the SOCs of the batteries are equal that the batteries release the same power to the grid-side converter. Hence, with different wind speeds, the GSC of WECS1, which receives larger average power from the wind turbine, will transfer a little more power to the utility grid. By contrast, the output power of WECS3 is lower compared to the others because the average wind speed flowing to WECS3 is the smallest. The battery power of both cases is similar because the average wind speed of WECSs is not much different, as depicted in Figures 17 and 18.

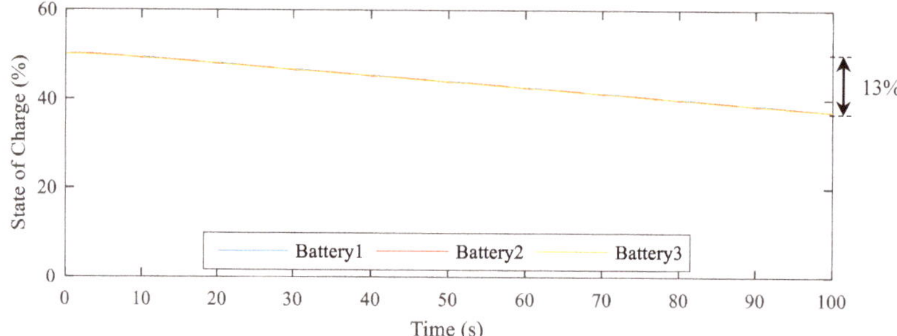

Figure 14. State of charge of batteries in the WECS with proposed controller (Case 2).

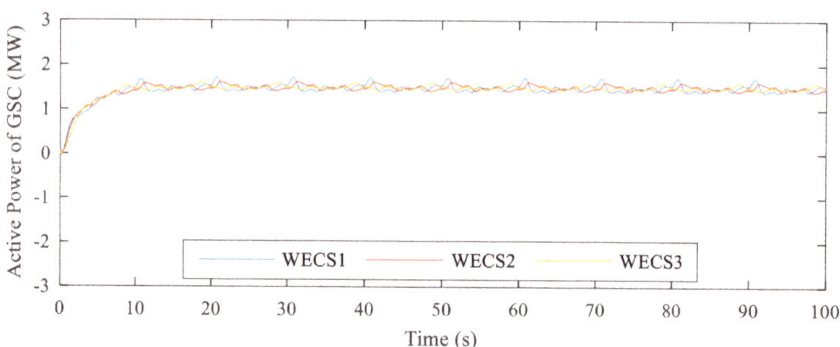

Figure 15. Active power of grid-side converter without proposed controller (Case 2).

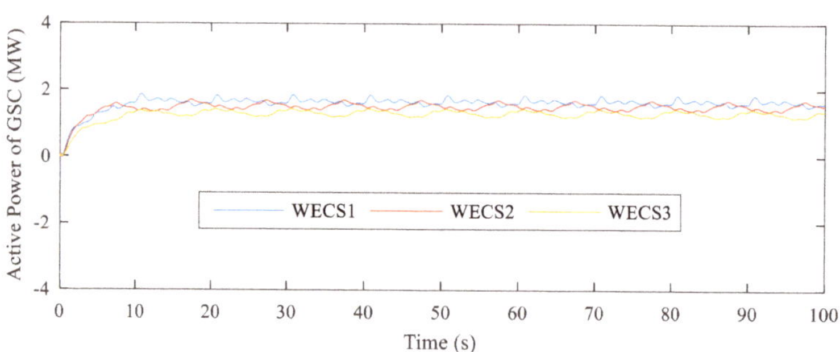

Figure 16. Active power of GSC with proposed controller (Case 2).

Furthermore, Figure 19 illustrates the total power of GSCs flowing into the utility grid in the case of a different initial SOC. The power fluctuation is smoothed because the fluctuation in wind power caused by the wind speed is avoided by means of the low pass filter in the GSC controller. There is a slight difference in the total power of the wind farm in the cases of with and without the proposed controller since two controllers consist of the strategy of smoothing wind power fluctuation.

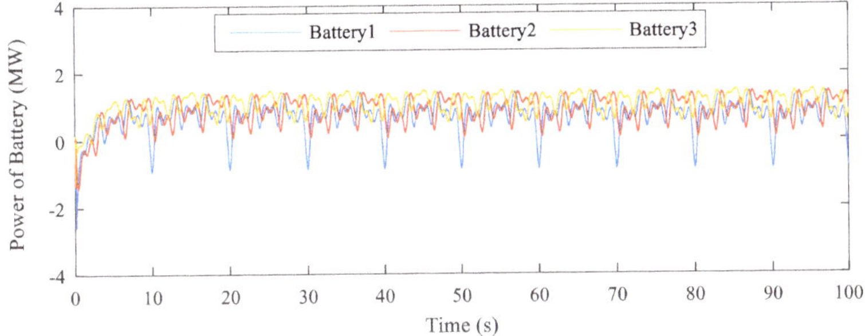

Figure 17. Power of battery without proposed controller (Case 2).

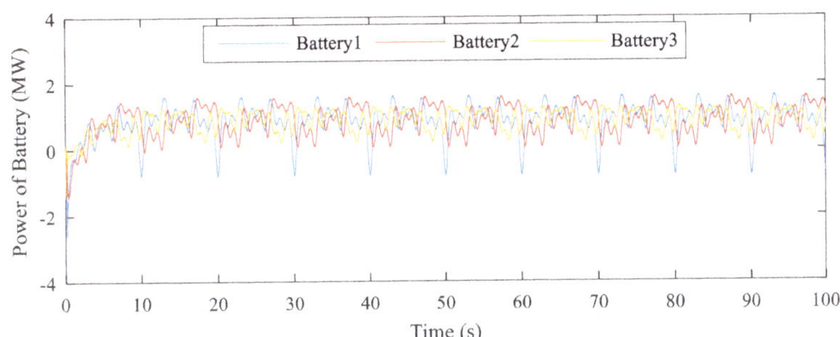

Figure 18. Power of battery with proposed controller (Case 2).

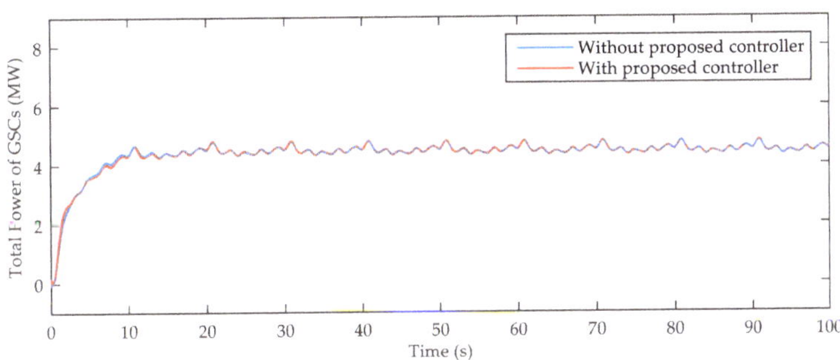

Figure 19. Total active power of GSCs in case of different initial SOCs.

5. Conclusions

In this paper, an improved VSG control based on consensus algorithm has been applied to the GSC for balancing the SOCs of batteries in the WECSs. The effectiveness of the proposed controller has been verified through different initial SOCs and same initial SOC conditions under different wind speeds. With the consensus-based SOC balancing controller, the batteries which have higher SOC could support the batteries which have lower SOC. Thus, this prevented the negative effects of the SOC imbalance on the operation of the WECS. In addition, the smoothing power control based on low pass filter has been applied to the GSC for compensating the active power fluctuation to the utility grid. Hence, beside the advantage of VSG to enhance the inertia to the WECS, the output power of the WECS has been smoother, while the SOC of batteries has been balanced.

Author Contributions: C.-K.N. conceived and designed the experiments; T.-T.N. and H.-J.Y. performed the experiments and analyzed the results; H.-M.K. revised and analyzed the results; C.-K.N. wrote the paper.

Funding: This research was supported by Korea Electric Power Corporation. (Grant number: R18XA03.)

Conflicts of Interest: The authors declare no conflict of interest.

References

1. Blaabjerg, F.; Teodorescu, R.; Liserre, M.; Timbus, A.V. Overview of control and grid synchronization for distributed power generation systems. *IEEE Trans. Ind. Electron.* **2006**, *53*, 1398–1409. [CrossRef]
2. Bui, V.H.; Hussain, A.; Kim, H.M. A multiagent-based hierarchical energy management strategy for multi-microgrids considering adjustable power and demand response. *IEEE Trans. Smart Grid* **2018**, *9*, 1949–3053. [CrossRef]
3. Nguyen, T.T.; Yoo, H.J.; Kim, H.M. A Droop Frequency Control for Maintaining Different Frequency Qualities in Stand-Alone Multi-Microgrid System. *IEEE Trans. Sustain. Energy* **2018**, *9*, 599–609. [CrossRef]
4. Johnson, L.P.K. Control of wind turbines. Approaches, challenges, and recent developments. *IEEE Control Syst. Mag.* **2011**, *58*, 44–62.
5. Luo, C.; Ooi, B.T. Frequency deviation of thermal power plants due to wind farms. *IEEE Trans. Energy Convers.* **2006**, *21*, 708–716. [CrossRef]
6. Gautam, D.; Vittal, V.; Harbour, T. Impact of increased penetration of DFIG-based wind turbine generators on transient and small signal stability of power systems. *IEEE Trans. Power Syst.* **2009**, *24*, 1426–1434. [CrossRef]
7. Nguyen, C.L.; Lee, H.H. A novel dual-battery energy storage system for wind power applications. *IEEE Trans. Ind. Electron.* **2016**, *63*, 6136–6147. [CrossRef]
8. Ganti, V.C.; Singh, B.; Aggarwal, S.K.; Kandpal, T.C. DFIG-based wind power conversion with grid power levelling for reduced gusts. *IEEE Trans. Sustain. Energy* **2012**, *3*, 12–20. [CrossRef]
9. Luo, F.; Meng, K.; Dong, Z.Y.; Zheng, Y.; Chen, Y.; Wong, K.P. Coordinated operational planning for wind farm with battery energy storage system. *IEEE Trans. Sustain. Energy* **2015**, *6*, 253–262. [CrossRef]
10. Han, W.; Zou, C.; Zhou, C.; Zhang, L. Estimation of cell SOC evolution and system performance in module-based battery charge equalization systems. *IEEE Trans. Smart Grid* **2018**. [CrossRef]
11. Han, W.; Zhang, L. Mathematical analysis and coordinated current allocation control in battery power module systems. *J. Power Sources* **2017**, *372*, 166–179. [CrossRef]
12. Nguyen, C.L.; Lee, H.H.; Chun, T.W. Cost-optimized battery capacity and short-term power dispatch control for wind farm. *IEEE Trans. Ind. Appl.* **2015**, *51*, 595–606. [CrossRef]
13. Wu, Q.; Guan, R.; Sun, X.; Wang, Y.; Li, X. SoC Balancing Strategy for Multiple Energy Storage Units with Different Capacities in Islanded Microgrids Based on Droop Control. *IEEE J. Emerg. Sel. Top. Power Electron.* **2018**, *6*, 1932–1941. [CrossRef]
14. Manenti, A.; Abba, A.; Merati, A.; Savaresi, S.M.; Geraci, A. A new BMS architecture based on cell redundancy. *IEEE Trans. Ind. Electron.* **2011**, *58*, 4314–4322. [CrossRef]
15. Lu, X.; Sun, K.; Guerrero, J.M.; Vasquez, J.C.; Huang, L. Double-quadrant state-of-charge-based droop control method for distributed energy storage systems in autonomous DC microgrids. *IEEE Trans. Smart Grid* **2015**, *6*, 147–157. [CrossRef]
16. Lu, X.; Sun, K.; Guerrero, J.M.; Vasquez, J.C.; Huang, L. State-of-charge balance using adaptive droop control for distributed energy storage systems in DC microgrid applications. *IEEE Trans. Ind. Electron.* **2014**, *61*, 2804–2815. [CrossRef]
17. Sun, X.; Hao, Y.; Wu, Q.; Guo, X.; Wang, B. A multifunctional and wireless droop control for distributed energy storage units in islanded AC microgrid applications. *IEEE Trans. Power Electron.* **2017**, *32*, 736–751. [CrossRef]
18. Maharjan, L.; Inoue, S.; Akagi, H.; Asakura, J. State-of-charge (SOC)-balancing control of a battery energy storage system based on a cascade PWM converter. *IEEE Trans. Power Electron.* **2009**, *24*, 1628–1636. [CrossRef]
19. Guan, Y.; Vasquez, J.C.; Guerrero, J.M. Coordinated secondary control for balanced discharge rate of energy storage system in islanded AC microgrids. *IEEE Trans. Ind. Appl.* **2016**, *52*, 5019–5028. [CrossRef]
20. Sahoo, S.K.; Sinha, A.K.; Kishore, N.K. Control Techniques in AC, DC, and Hybrid AC–DC Microgrid: A Review. *IEEE Trans. Emerg. Sel. Top. Power Electron.* **2018**, *6*, 738–759. [CrossRef]

21. Li, C.; Coelho, E.A.A.; Dragicevic, T.; Guerrero, J.M.; Vasquez, C. Multiagent-based distributed state of charge balancing control for distributed energy storage units in AC microgrids. *IEEE Trans. Ind. Appl.* **2017**, *53*, 2369–2381. [CrossRef]
22. Guan, Y.; Meng, L.; Li, C.; Vasquez, J.; Guerrero, J. A dynamic consensus algorithm to adjust virtual impedance loops for discharge rate balancing of ac microgrid energy storage units. *IEEE Trans. Smart Grid* **2017**. [CrossRef]
23. Rocabert, J.; Luna, A.; Blaabjerg, F.; Rodriguez, P. Control of power converters in AC microgrids. *IEEE Trans. Power Electron.* **2012**, *27*, 4734–4749. [CrossRef]
24. Zhong, Q.C.; Weiss, G. Synchronverters: Inverters that mimic synchronous generators'. *IEEE Trans. Ind. Electron.* **2011**, *58*, 1259–1267. [CrossRef]
25. Nguyen, C.K.; Nguyen, T.T.; Yoo, H.J.; Kim, H.M. Improving Transient Response of Power Converter in a Stand-Alone Microgrid Using Virtual Synchronous Generator. *Energies* **2017**, *11*, 27. [CrossRef]
26. Pogaku, N.; Prodanovic, M.; Green, T.C. Modeling, analysis and testing of autonomous operation of an inverter-based microgrid. *IEEE Trans. Power Electron.* **2007**, *22*, 2. [CrossRef]
27. Olfati-Saber, R.; Fax, J.A.; Murray, R.M. Consensus and cooperation in networked multi-agent systems. *Proc. IEEE* **2007**, *95*, 215–233. [CrossRef]

MDPI

St. Alban-Anlage 66

4052 Basel

Switzerland

Tel. +41 61 683 77 34

Fax +41 61 302 89 18

www.mdpi.com

Energies Editorial Office

E-mail: energies@mdpi.com

www.mdpi.com/journal/energies